White Identities: Historical and International Perspectives

Alastair Bonnett

PRENTICE HALL

An imprint of PEARSON EDUCATION

Harlow, England · London · New York · Reading, Massachusetts · San Francisco · Toronto · Don Mills, Ontario · Sydney
Tokyo · Singapore · Hong Kong · Seoul · Taipei · Cape Town · Madrid · Mexico City · Amsterdam · Munich · Paris · Milan

Pearson Education Limited
Edinburgh Gate
Harlow
Essex CM20 2JE
England
and Associated Companies throughout the world

Visit us on the world wide web at:
http://www.pearsoned-ema.com

First published 2000

ISBN 0 582 35627 X

British Library Cataloguing-in-Publication Data
A catalogue record for this book is available from the British Library

Library of Congress Cataloging-in-Publication Data
A catalog record for this book is available from the Library of Congress

Typeset by 42 in 10/11pt Palatino
Produced by Addison Wesley Longman Singapore (Pte) Ltd.
Printed in Singapore

For Louis

Contents

Contents

List of Figures

Acknowledgements

Chapter 1 is a revised version of an article previously published in *Ethnic and Racial Studies*, **21**, 6, 1998.

Chapter 2 is a revised version of an article previously published in *The Journal of Historical Sociology*, **11**, 3, 1998.

Chapter 4 incorporates material previously published as 'The new primitives: identity, landscape and cultural appropriation in the mytho-poetic men's movement', in *Antipode*, **28**, 3, 1996.

Chapter 5 incorporates material previously published as 'White identities and the critique of anti-racism', *New Community*, **22**, 1, 1996, as 'Constructions of whiteness in European and American anti-racism', in P. Werbner and T. Modood (eds) (1997) *Debating Cultural Hybridity: Multi-cultural Identities and the Politics of Anti-racism* (London, Zed Books), as 'White Studies: the problems and possibilities of a new research agenda', *Theory, Culture and Society*, **13**, 2, 1996 and as 'Review symposium', *Race, Ethnicity and Education*, **1**, 2, 1998.

Thanks to Matthew Smith and Shuet-Kei Cheung at Pearson Education for their patience and support for this project. Thanks also to Richard Collier, Rachel Holland, Nina Laurie, Anoop Nayak and Neil Ward for their help and advice.

We are grateful to the British Library for permission to reproduce Figure 1.1 from the Ms. of the Khamsa of Nizami, Isfahan, dated 1076–7/1665–7. ADD6613, fol. 208a.

Introduction

All this whiteness that burns me ...
(Fanon, 1986, p.114)

This book is about the relationship between white identity and modernity. It explores the development of whiteness as a social ideal across the planet and the way different groups have been affected by this process.

The book is not all-inclusive. The subject is just too large and too important to make such a claim. Much has been left out. However, what remains should serve as an introduction: readers are offered a basic narrative of the global history of racial whiteness, with a few episodes explored in some detail.

I shall start, though, with a complaint (perhaps it is more of an injured whine). It has not always been easy writing a book about a subject that most people think they know all about already. The reaction of white people to my endeavours has often been bemusement, tinged with annoyance. 'I know about that already,' I have been informed on more then one occasion, with the added explanation, 'I *am* white, so what's to know?' People who aren't white sometimes also asserted an intimate familiarity with the topic: they knew all about it too, because 'it's a white world, isn't it?' and, as I've been told a few times, 'aren't most books about white people anyway?' Even amongst the *cognoscenti* of ethnic and racial studies, a sense of familiarity prevails. Whiteness has been consigned to an emergent specialism, 'white studies', a sub-field that has, with a handful of exceptions, arisen from and limits its horizon of interest to the United States of America. It is now a known quantity, an example of the sophisticated state of the 'race debate' in the USA for supporters, an indication of its ridiculous 'political correctness' to its many detractors.

To have one's explorations of an area dismissed as common knowledge before they have even been committed to print is a curious sensation. It is a sense compounded by my suspicion that people actually like the idea that there is nothing much to say about whiteness; that to discuss the subject is either impossible or an entirely rhetorical exercise. It is against this claustrophobic background, in the face of this all-knowing audience, that

1

I issue a simple claim: *this book will change the way you look at 'the white race'*. It will do so, I suspect, largely because it will provide information on the topic that you did not already know. Drawing on material from ancient China to contemporary Brazil, from the West and the non-West, the book offers a series of views on the emergence and impact of whiteness that, although far from comprehensive, demonstrate both the global dimensions and the historical depth of the subject. It will show that the emergence of white racial identities is an integral component of the development of modernity across the world. Indeed, it is my contention that one cannot grasp the development of the modern world, and more especially the notion of what is modern and what is not, without an appreciation of the racialised nature of modernity, and, more particularly, of its association with a European identified white race.

My approach is introductory in both its scope and method. Five key aspects of white identities are addressed: the development of racial whiteness; the particularity of its local forms; its impact outside the West; the attempt to escape whiteness; and, finally, the relationship between whiteness and anti-racism. The book offers the widest possible overview of white identities and does so without presuming that readers are acquainted with either 'white studies' or any of the manifold academic disciplines that I have found it necessary to draw upon. As this implies, I have not provided a collection of summaries of what other authors have written about white identity, or sought to offer a detailed introduction to white studies (although Chapter 5 goes some way to providing the latter). Unfortunately, despite some invaluable contributions to the area, the historical limitations and narrow geographical focus of existing work in white studies means that it provides only a small window onto the subject. Thus this book will venture, every so often, onto new ground, into areas that have not been written about before, at least not with a focus on whiteness. Not unrelatedly – and to avoid disappointment – I should also add that I have decided not to spend a great deal of time examining the history of whiteness and race in the country that dominates the existing literature on these topics, the USA. Hence, readers interested solely in whiteness in the latter society might be better advised to look elsewhere (I would recommend Jacobson, 1998; Roediger, 1992; 1994; 1998; Morrison, 1992; Hill, 1997; and Ignatiev, 1995).

Perhaps I should also confess that I have sought to avoid offering too many generalisations derived from my own personal experiences as a white man as source material to pronounce on the world at large. Since reflexivity understood as autobiography has come to be considered the ultimate in enlightened conduct within social studies this abnegation of my responsibility to situate myself in relation to my subject may appear perverse. However, I don't think it is coincidental that the most valuable contributions to white studies by white people, the ones that have actually moved the debate forward, are those that are least self-absorbed (I am thinking here of Dyer, 1997, and the US historians Roediger, Jacobson and Ignatiev[1]). By contrast, the endless musings and reminiscences that characterise an increasing number of engagements with the issue (for example, Wray and Newitz, 1997; see also Fine *et al.*, 1997; Thompson and

Tyagi, 1996) may offer academics alluring opportunities to write about themselves but provide little context or insight into the social formation of whiteness. It is worth pointing out that white people, white men in particular, have long had the power and, evidently, the inclination to drone on about themselves. In his analysis of the cultural imagery of whiteness, Richard Dyer (1997) highlights this danger. 'Writing about whiteness', he notes, 'gives white people the go-ahead to write and talk about what in any case we have always talked about: ourselves' (p.10). Dyer also suggests that many white people appear keen to talk about themselves in the same kind of way as other groups. In other words, that in the context of 'everyone else' being able to to *assert themselves* racially and ethnically, some white people desire to do the same, to indulge and explore their 'racial essence'. Dyer calls this,

> the problem of "me-too-ism", a feeling that, amid all this (*all* this) attention being given to non-white subjects, white people are being left out. One version of this is simply the desire to have attention paid to one, which for whites is really only the wish to have all the attention once again. (p.10)

Such agendas, far from producing work that sheds light on white identities, will only help to further reify the subject. I would submit that the best way of avoiding such narcissistic, self-celebratory accounts is to push the historical and geographical boundaries of our understanding of whiteness as far as we can. In order to disorient and subvert the complacencies of racial cliché built up within one's own society and one's own experiences, one needs to learn and think about places and peoples from other times, other places. For some, such global ambitions may have a slightly colonial ring to them. After all, from the 1950s onwards, ethnic and racial studies in the West have travelled in precisely the opposite direction, rejecting imperialist anthropology for the dissection of racism in Western nations. Whilst I am in agreement with the political motivations behind this project, it has had the effect of making debates on race in the West seem absurdly insular. Moreover, the coherence of abstracting particular national narratives from the world economy – of studying 'French racism' or 'British anti-racism' – is increasingly questionable. There is now a need for writers in this field to broaden their horizons. White identities are, if nothing else, global phenomena, with global impacts. Indeed, the nature and implications of their local manifestations only come into view when they are understood *as local*. As this implies, I am not advocating a return to regional specialisms (the Africanist, the Sinologist and so on), but an attempt to engage the international and comparative diversity of whiteness. It is an approach vulnerable to eclecticism, to pirouetting from Peru to Poland in a way that inevitably infuriates readers brought up to respect national boundaries. But this is a risk worth taking, a risk that must be taken if we are ever to think and act in an informed way about racial prejudice or whiteness.

Since this introduction seems to have turned into a list of 'risks and 'dangers', let me conclude with one last warning. It is not my intention in this book to suggest that whiteness is necessarily a determining factor in each and every social or political situation. Although integral to modernity, its form and importance are historically and geographically contingent. The confusion and conflation of whiteness with words that at different times and in different places have been associated with it, such as 'European' and 'Western', necessitate that the term's synonyms, its cognitive alignments, form part of our study. Moreover, as other authors have explained, whiteness is always enacted in association with class and gender (Frankenberg, 1993; Roediger, 1992; 1994). I would hope this text does not inspire anyone to begin claiming that whiteness is always and everywhere the most pressing issue of the hour. The utility of the present study lies in its insistence upon the context of whiteness, its place *within a wider system of social change*. To fetishise whiteness is to miss the point. Moreover, as I hope to show, the power of whiteness rarely works in such as a way as to allow one to isolate and dispatch a discrete 'bad racist white influence'. The uncomfortable truth is that white identity resides in social forces and categories, such as 'modernisation', 'development' and 'civilisation', with which we are all engaged in some way. We may justifiably hope to encourage the deracialisation of these processes, to open them up to less Eurocentric ends. Whiteness can, and I am in no doubt one day will, be superseded, made to appear as archaic an identity as Tuton or Gaul. But to imagine that it is a *homogenous and alien* 'enemy' – something other to, or inherently outside of, anti-racism – is a romantic delusion.

Introducing white identities in five chapters

My focus on the social formation of whiteness is methodologically narrowed in the following five chapters by a concentration on the way people categorise and represent themselves and others. I shall be drawing on a variety of first-hand testimonies, from writers, politicians and other commentators, to explain and to exemplify the nature of white identity.

The book starts with a broad overview of the rise of racial whiteness. More specifically, Chapter 1 provides a critical history of the conflation of European and white identities. It commences with a discussion of premodern white identities in China and the Middle East. The production of a racialised European white identity is then examined. The obsessional, exclusionary character of European racial whiteness is related to the gradual marginalisation of non-European white identities. Drawing primarily on late nineteenth-century and early twentieth-century commentaries, it is also argued that the excessive idealisation of whiteness characteristic of its modern European form engenders an unstable and contradictory identity: in societies structured upon class, ethnic and gender hierarchies the 'burden of whiteness' cannot be equally apportioned.

Chapter 2 offers a geographically local view of the formation of white identity. In contrast to the previous chapter, it may be seen as a case study of the development of racial whiteness within one particular country. More precisely, the chapter offers an explanation of how and why the

British working class, from being marginal to white identity in the nine-teenth century, came to adopt and adapt this identity in the twentieth century. The changing position of whiteness within the symbolic consti-tution of capitalism is discussed. The transition from whiteness as a bour-geois identity within Victorian, relatively laissez-faire, capitalism to whiteness as a popular, or mass, identity within the more state interven-tionist capitalism of the twentieth century is used to exemplify the muta-ble nature, and political complexities, of the relationship between working class and white identities.

Chapter 3 opens out the geographical focus again to explore the forma-tion of 'white modernity' in non-Western societies in the twentieth cen-tury. The empirical emphasis is upon Latin America and Japan. My account is divided into two. In the first section I address Latin America, focusing on the history of 'national whitening' through selective immi-gration policies in Brazil and Venezuela, before turning to a very specific contemporary phenomenon, namely the appeal of the white, blonde Bra-zilian 'megastar' Xuxa. In the second section I introduce material from East Asia. More specifically, I discuss the history of the adoption but also the assimilation – what might be called the strategic reading and appro-priation – of whiteness in contemporary Japan.

It is no easy thing to escape whiteness, whatever one's skin colour. To invert the expectations and ideals of whiteness can all too easily slide into appropriating a white, not to mention colonial, myth of the savage and exotic other. This may be framed as 'the problem of primitivism'. It is a problem most commonly seen as visited upon white radicals who feel themselves alienated from 'white, Western, capitalist' society. However, as Chapter 4 argues, primitivism is neither the preserve of a radical elite nor of white people. It is part and parcel, an inevitable consequence, of the development and dissemination of racialised modernity and, as such, a socially ubiquitous and international force. Despite this slightly unor-thodox thesis, Chapter 4 also aims to provide a conventional introduction to the subject of primitivism. It has four sections. The first is devoted to the primitivism of the group usually associated with the term 'Western avant-garde artists'. The second section concerns a particular individual, Grey Owl (also known as Archie Belaney), an Englishman who not only chose to live as a native Canadian but who also gave lectures throughout England and North America in the 1930s (preaching his message of eco-logical conservation and anti-modernity), whilst claiming to be a 'half-caste'. The mythopoetic men's movement forms the empirical focus of the third section of Chapter 4. Like many primitivists, mythopoetic men have been subjected to a lot of ridicule and, not unrelatedly, seem to transgress many expectations concerning the conduct of modern, civilised adults. They provide a particularly explicit example of the intersection of mascu-linity and primitivism – more precisely, the way escapes from whiteness can be conflated with a flight into the supposedly more natural and stable gender relations of pre- or non-modern societies. The class dynamic of the movement is also addressed, with the widespread notion that it reflects a current 'crisis of middle-class masculinity' being, if not disproved, at least found sociologically simplistic. The final section of

Chapter 4 looks, albeit it in an eclectic and unsystematic fashion, at the place of primitivism in non-Western societies. More especially, the relationship between primitivism, the development of national identity and the promotion of tourism is discussed in a number of so-called 'developing' countries.

The fifth and final chapter addresses the relationship between white identity and anti-racism. The image and role of whites in anti-racism is explored. Drawing on British and North American material, it is suggested that white identity is reified within most orthodox anti-racism, a process that offers whites the position of altruistic observers within the anti-racist struggle, of people whose good will must be constantly appealed to but who have no real stake in anti-racist change. Against this, some more recent and nuanced tendencies within anti-racist 'white studies' are considered.

---

1.   The two exceptions I would make to this rather blanket observation are Vron Ware's (1992) study of the emergence of 'white feminism', *Beyond the Pale: White Women, Racism and History*, and·Crispin Sartwell's (1998) painfully self-conscious but subtle observations on seeing his own whiteness in the mirror of African-American identity, *Act Like You Know: African-American Autobiography and White Identity*.

CHAPTER 1

Who was white? The disappearance of non-European white identities and the formation of European racial whiteness

Introduction

The topic of white identity is currently in vogue within racial and ethnic studies. Most significantly, the nineteenth- and twentieth-century history of how different European groups became accepted as white in the United States has begun to be mapped, principally by US historians (Roediger, 1992; 1994; Ignatiev, 1995; Jacobson, 1998). Yet this, otherwise insightful, critical moment still remains partly trapped inside the mythologies of European whiteness. For whilst European, relatively recent, forms of whiteness are attracting attention, other experiences of whiteness, developed before the late modern era or outside North America and Europe, have receded ever further from view.

The historically and geographically narrow focus of the current debate on whiteness means that the particularity of its modern form continues to evade analysis. Only by positioning European-identified and racialised whiteness within a longer and broader view of white identities can the power of European societies to assert and insert their social categories and symbolisms across the globe be properly understood.

This chapter provides a critical history of the Europeanness and racialisation of whiteness. It advances the necessity of a longer historical, and wider geographical, view of the production of white identities and a more sceptical attitude towards the stability of its European configurations. More specifically, it seeks to examine two interrelated historical processes:

1. The development of non-European (and non-racialised) white identities and their marginalisation or erasure by an increasingly hegemonic, European-identified, racialised whiteness. It may be useful to mention, at the outset, a central irony of this process: for whilst the eighteenth and nineteenth centuries witnessed the development of an exclusionary association between being white and being European, a counter-tendency amongst racial scientists encouraged the application of the term 'white' to a much wider set of people. As we shall see, this latter,

'scientific', approach has itself been marginalised by the former, more popularist, discourse.

2. It will be argued that the development of whiteness as a racialised, fetishised and exclusively European attribute produced a contradictory, crisis-prone, identity. Two sets of conflicting discourses are implicated in this process: first, colonial, imperial and national rhetorics of European racial equivalence that, ostensibly, offered the privileges of white identity to all European-heritage peoples; second, the denial or marginalisation of certain European-heritage groups' whiteness, a process of racial suspicion fostered by social exclusions based on gender, class and ethnicity.

These two arguments are developed in three sections. The chapter commences with a historical account of the formation of non-European white identities. This discussion draws primarily on examples from China and the Middle East. This account is framed by a discussion of the problematic nature of 'racial histories' of pre-modern and non-European societies. The second section advances an assessment of the marginalisation or erasure of Chinese and Middle Eastern white identities and the formation of a European racialised whiteness. Third, the ethnic, class and gender particularities and contradictions of this latter identity are examined.

Pre-modern white identities: China and the Middle East

The modern idea of 'race' is distinctive because it emerged from modern attitudes towards nature and politics. In other words, it is the product of European naturalist science and European colonial and imperial power (Miles, 1989; Guillaumin, 1995). If we accept this formulation then we must also conclude that the terminology and praxes of biologically rationalised racial discrimination could not have existed in pre-modern societies. A related implication is that to identify evidence of a population group which defined itself, partly or wholly, by a term that is commonly translated into English as 'white' is not necessarily to identify a racial or, indeed, an ethnic, history. Any attempt to chart 'a history of white identity' by imaginatively projecting onto other eras its modern form merely contributes to the latter's naturalisation and, hence, mystification. It follows that the history of the Europeanisation of whiteness is not a history of a European seizure of a pre-formed identity but, rather, a narrative of the ability to marginalise and forget other forms of white identity and to create, assert and disseminate a particular vision of human difference.

The tendency of scholarship into non-European colour consciousness to interpret the history of 'others' as merely a series of faltering reflections of the events and ideas of European society means that, before proceeding to my evidence for the development of white identities in China and the Middle East, some contextual remarks are required on the Western historical works that appear to provide most insights into these identities. The most important (and controversial) of these sources are Frank Dikötter's (1992) *The Discourse of Race in Modern China* and Bernard Lewis's (1971) *Race and Color in Islam*. As their titles imply, both works claim to have uncovered histories of racial thinking in, respectively, China and Islam (a term which Lewis translates geographically as the Middle East). Moreover, they assert that this form of thinking developed many centuries before the existence of Western racial science.

Dikötter is the most bold in this claim (see also Dikötter, 1990; 1994; 1997). He adapts Banton's (1987) typology of European racial thinking to China (chapter titles include, 'Race as lineage', 'Race as type' and 'Race as species') and asserts the virulence of 'racial consciousness' (p.2) and 'racial discrimination' (p.3) amongst elite groups in both ancient and medieval Chinese society (see also Kong, 1995). Crucially, he 'translate[s] by "race" (*zu, zhong, zulei, minzu, zhongzu, renzhong*, in Chinese) terms that appear to stress the biological rather than the sociocultural aspects of different peoples' (pp.viii–ix). The application of these translations to pre-modern material is highly problematic (Dirlik, 1993; Stafford, 1993). It should be noted, first, that at least one of these terms is a modern neologism. As Crossley (1990, p.19) notes, the word '*minzu*' 'has not yet been traced earlier than 1895'. More fundamentally, the term which Dikötter privileges as a synonym for 'race', '*zu*', 'evolved from a non-ascriptive, non-taxonomic word of generalized meaning to an ascriptive, taxonomic word' (Crossley, 1990, p.20) only in the eighteenth century. Moreover, even at this latter date, it still referred essentially to 'established, historical peoples'; a far cry from naturalist notions of race.

Even on the evidence of Dikötter's own citations of pre-modern origin his employment of the language of race must be called into question. For they neither refer to, nor emerge from, a discourse concerned with the objective classification of natural differences. Thus, although Dikötter draws the ancient and medieval Chinese tradition of calling certain Chinese people white into his racialised schema, his empirical evidence rebels against his thesis. For what emerges from these sources is not a scientific but a self-consciously symbolic, mythopoetic, rhetoric of white identity. Hence, although we may concur that elite Chinese 'developed a white–black polarity at a very early stage ... and called their complexion white from the most ancient days' (p.10; see also Maspero, 1978), it must also be noted that Dikötter exemplifies this process, not with evidence of racial thinking but with the following poem (quoted from a collection of poems from the seventh century BC, the *Shijing*),

> Her fingers were like the blades of the young white grass;
> Her skin was like congealed ointment;
> Her neck was like the tree-grub;
> Her teeth were like melon-seeds;
> Her head cicada-like;
> Her eyebrows the silkworm moth

> (cited by Dikötter, 1992, p.10)

It is my contention that, although there were no white *racial* identities in pre-modern China, there were white identities. In other words, certain

Chinese people employed the category 'white' to help define which social collectivity they belonged to. The poem cited above is a not untypical example of the way physical evidence of whiteness was positively connoted in early Chinese society. Whiteness was associated with purity, sensitivity and beauty. Drawing on similar verses, Isaacs (1968, p.92) notes that a 'celebration' of white skin colour moves 'gracefully through endless reams of ancient Chinese poetry'.

Betraying the influence of a later, racialised, category of Chinese identity (i.e., the notion of a 'yellow race'), the French Sinologist Maspero (1978, p.11) opines that the ancient Chinese 'had that yellow tinge which they always characterised as white'. In the context of encounters with 'darker-skinned peoples' from the west and south of China, whiteness was used to distinguish Chinese from non-Chinese peoples. In his study of notions of the exotic during the Tang Dynasty (AD 618–906), Schafer (1963; see also Isaacs, 1968) locates a persistent tendency to describe the 'otherness' of individuals from Persia to Indonesia in terms of their blackness. A later example of the same tendency is to be found in the works of Zhang Xie, a Chinese geographer of the early Ming (1368–1644) period, who noted that, 'people in Malacca have a black skin, but some are white: these are Chinese' (cited by Dikötter, 1992, p.11).

Although whiteness was used to define and identify Chinese people, this attribute does not appear to have become fetishised to the exclusion of other physical traits. Thus notions of smell and hair colour were also integrated into notions of Chinese collective identity. Not unrelatedly, the modern idea that white is an objective category rather than a description (as seen, for example, in the notion that all Europeans are white no matter what their skin complexion) is also far less apparent in Chinese interpretations of whiteness. Thus assertions of Chinese whiteness did not imply that other peoples could not be as, or even more, white in appearance. For example, Crossley (1990, p.10), drawing examples from the late nineteenth century, notes that 'Non-Manchus were commonly convinced that Manchus could be distinguished from Chinese by racial physical traits, like flat heads, or a remarkably white skin.'

That the poem cited earlier concerns a princess may also alert us to the fact that whiteness was associated with membership of the elite. In his study of anti-African discrimination in China, Sullivan (1994, p.440) asserts that the 'Chinese, who perceived their own skin to be white until the early 20th century, considered individuals with lighter skin as having a higher social status than dark-skinned peoples.' Sautman (1994, p.427) offers a more precise example and explanation of this process: the 'most attractive man' in '[t]raditional Chinese culture,' he notes, 'was a "white-faced scholar" (*baimian shusheng*) whose freedom from manual labour at once implied a high status, potentially leisured life and light complexion.'

For a significant period of Chinese history, the social elite were defined in contrast to 'black-headed' or peasant people. Lun (1975) dates the first application of the term 'black-headed people' to the Chinese peasantry to the mid-Warring States period (403–221 BC). The term referred to peasants' supposed black complexion. '[T]he peasants', Lun (p.247) explains, 'were exploited laborers, and heavy labor in the open fields caused their faces to be burned swarthy by the sun.'

However, Lun's work also indicates that any attempt to impose a simple black/white chromatisation upon early Chinese class relations would be misplaced. He notes that the Qin Dynasty (221–207 BC) decree (issued in 221 BC) declaring that 'the name of the common people is changed to black-headed [people]', reflected a *rise* in social status for this group, more specifically a change from slave to peasant status. Futhermore, Lun (p.252) asserts that during the Qin Dynasty black 'became the color that symbolized the government ... and was the most exalted color'. This instance of the political mutability of colour symbolism is suggestive of the fact that whiteness, whilst valued amongst the Chinese elite, was understood as a social and political creation. In other words, whiteness was not naturalised as a 'commonsense' explanation, or even necessarily a core component, of elite groups' social superiority.

Early encounters with Europeans do not appear to have disturbed Chinese white identities. Westerners were not interpreted as more authentically white than Chinese people. Indeed, many accounts emphasise the peculiar, ash-like, quality of the former's skins. Thus whilst Zhang Xie depicted the Chinese as white he noted that the Portuguese 'are seven feet tall, have eyes like a cat, a mouth like an oriole, an ash-white face'(quoted by Dikötter, 1992, p.14). This interpretation survived into the nineteenth century, Europeans being described 'as cold and dull as the dead ashes of frogs' by Jin He (quoted by Issacs, 1968, p.91). A tradition of identifying Europeans not as white but as red was also maintained into the last century. For one nineteenth-century traveller Europeans were 'reddish purple', a people who 'greatly resemble the Mongols' (quoted by Dikötter, 1992, p.54).

A similar set of sentiments appears in medieval literature from the geographical area described in the contemporary West as the 'Middle East'. The term 'white' was routinely used to identify Middle Eastern peoples and distinguish them from darker-skinned others, a tradition that lingers on today. It has been argued that the formation of relatively stable colour-coded group identities occurred in the context of the expansion by Islamic peoples further into Asia and Africa from the eighth century (Lewis, 1971; Morabia, 1985; Bédoucha, 1982; Lindholm, 1996). Bédoucha (1982, p.533) notes that,

> From the death of the prophet, with the commencement of the conquests in Asia and Africa, the situation ... changed. This change is clear in the literature. References to nuances of colour that characterise individuals disappear to be replaced by three terms – white, red and black – that have clear ethnic meanings and which, in respect to the blacks, indicate inferiority.

However, Bédoucha also warns against taking this process out of its specific social and historical context. More specifically, he takes Lewis's book *Race and Color in Islam* to task for comparing pre-modern Middle Eastern societies with South African apartheid and Nazi Germany. Reflecting another strand of criticism, Nyang and Abed-Rabbo (1984, p.267) comment that *Race and Color in Islam* was written to provide 'ammunition to those anti-Arab and anti-Islamic elements in the West' and that its charges of racism are examples 'of the European pot calling the Arab kettle black'. In fact, Lewis's misapplication of racial categories is little different from Dikötter's. Both seek to use Western, modern, categories to understand non-Western, pre-modern societies.

Despite this problematic approach, the work of Lewis and Dikötter evidences a wealth of research of direct relevance to any serious study of identity formation in China and the Middle East. For, although 'racism' could not have existed in ancient or medieval Middle Eastern societies, Lewis provides clear evidence that colour-coded identities and social discriminations certainly did.

Like Bédoucha, Lewis argues that a 'specialization and fixing of color terms' followed 'the great Islamic conquests' in Asia and Africa (p.9; see also Lewis, 1990). One of the ways he evidences this development is by including in *Race and Color in Islam* numerous painted illustrations taken from a range of Middle Eastern and North African manuscripts. All of these images appear to evidence colour-consciousness, members of the social elite being rendered as white whilst slaves, servants and other groups tend to be painted black or brown (Figure 1.1 shows one of the seventeenth-century manuscripts used by Lewis). However, it must also be noted that Lewis offers no commentary on these images; they are allowed to 'speak for themselves', a manoeuvre than inevitably lends itself to the kind of decontextualised, modern reading of source material objected to by Bédoucha. Lewis's use of textual evidence is more convincing. In particular, he draws attention to the ninth-century writer Jahiz of Basra's essay 'The Boast of the Blacks Against the Whites'. The essay is a defence of 'the Blacks' against charges made by 'the Whites'. Clearly conversant with a set of colour-coded stereotypes of group identity, Jahiz argued that whites did not have a monopoly on either intelligence or beauty. (Lewis concludes, however, that, on the basis of his other, 'anti-Black' work, this defence was satirical.) The same colour-coded set of identities was employed by the group of Middle Eastern poets of African origin known as 'the crows of the Arabs'. Nusayb (quoted by Lewis, 1971, p.12), one such poet, writing in the eighth century, attempted to challenge prevalent stereotypes with the following observation:

> Some are raised up by means of their lineage; the
> verses of my poems are my lineage!
> How much better a keen-minded, clear-spoken
> black than a mute white!

There is also evidence to suggest that, as in China, a white complexion was associated with membership of the social elite. In support of this thesis, Lewis (1971, p.95) cites the eleventh-century Tunisian poet Ibn Rashiq's celebration of the city of Qayrawan: 'How many nobles and gentlemen (*sayyid*) were in it,/with white faces'. In his historical anthropology of the Middle East, Lindholm (1996, p.222) takes a similar position, noting that 'color, [was] used by urbanites, villagers and tribesmen alike' as one of the categories 'that could legitimize inferiority'.

Figure 1.1 'Black servant in attendance' (original caption), from B. Lewis, *Race and Color in Islam* (1971)

In another parallel with the Chinese experience, it appears that early contact with 'Europeans' did not destabilise Middle Eastern peoples' white identities. Indeed, a range of colours was applied to 'Europeans', including white but also yellow, red and pale blue. For the tenth-century geographer Masudi (cited by Lewis, 1982, p.139), peoples from the area we now call Europe are 'so excessively white that they look blue; their skin is fine and their flesh coarse. Their eyes, too, are pale blue, matching their coloring.'

In both pre-modern China and the Middle East, whiteness was a valued physical attribute. Not unrelatedly, in both societies whiteness was incorporated into dominant groups' collective identities. In neither society does contact with 'Europeans' appear to have disturbed this process. However, in neither society do we find the kind of fetishisation of whiteness, its use as a central icon of identity, later evident amongst Europeans. The positive connotations of whiteness were not reified into a natural attribute.

The association of whiteness with positive qualities was far from being universal in pre-modern societies. Moreover, in many societies whiteness was embroiled in more than one set of connotative traditions. In China, for example, as in many other societies (both pre-modern and modern), whiteness was (and is) seen as the colour of death and mourning. Similar traditions exist in South America and Africa (Chevalier and Gheerbrant, 1996). Describing twentieth-century Kongo cosmology, MacGaffey (1994, p.255) explains that the

> dead … contrast sharply with the living in some respects, one of which is that they are white in color … This same whiteness, contrasting with the organic and domestic blackness of charcoal, appears in masks all over Central Africa.

A similar example is offered by Robert Harms (1981) in his study of identity constructions amongst the peoples of the central Zaire basin in the nineteenth century. Harms (p.210) notes that, 'White people were … associated with spirits of dead ancestors … Indeed, Mpoto, the name generally taken to mean "the country where white people came from", actually means "the land of the dead".'

Whether positively or negatively connoted, whiteness was widely employed within the identity constructs of non-European and pre-modern societies. All these forms of whiteness have now been either forgotten or marginalised. When a contemporary scholar, such as Harms, in the citation above, refers to 'white people' *he does not have to explain* to whom he is referring. Today, with certain limited and increasingly residual exceptions, the term 'white' is equated with the term 'European'. No other interpretation is deemed possible (not, at least, within the literature of the West). The next section investigates how this happened.

From white to non-white: a modern disappearance

We have seen that non-Europeans' contact with Europeans did not necessarily disturb the former's white identities. My first task in this section will be to show that early modern European accounts of contact evidence few examples of a concern to contradict such interpretations or assert Europeans' sole claim to whiteness. As this implies, such encounters should not necessarily be interpreted as 'key moments' in the establishment of an exclusionary and hegemonic European white identity (cf. Cox, 1948; Malcomson, 1991). As we shall see, the latter process, with its concomitant marginalisation of other forms of whiteness, occurred significantly later.

Contacts with white non-Europeans

Accounts of early European travellers encountering 'white people' in non-European lands are numerous. Thus, for example, we find, as a study by Reid (1994, pp.274–5) shows, that 'Portuguese conquistadors routinely described their Gujerati or Arab antagonists as white, as well as the Chinese and Ryukyuans.' The first European mission to the Qing area of China described the inhabitants as having a white appearance 'equal to the Europeans' (cited by Dikötter, 1992, p.55). The OED (1989) offers another example from a travel book published in 1604. Grimstone's translation of Acosta's *The Natural and Morall Historie of the East and West Indies* describes 'a part of Peru, and of the new kingdom of Grenado, which ... are very temperate Countries ... and the inhabitants are white.' However, the fullest account I have read of European recognition of whiteness in others is by Vaughan (1995) in his account of early settler representations of Native Americans (see also Epperson, 1997). In summary, Vaughan shows that

> not until the middle of the eighteenth century did most Anglo-Americans view Indians as significantly different in color from themselves, and not until the nineteenth century did red become the universally accepted color label for American Indians. (p.4)

Thus, for example, George Best (quoted by Vaughan, 1995, p.9), writing in 1578, noted 'under the Equinoctiall in *America* the people are not blacke, but white.' This interpretation appears to have been assisted by the widespread conviction that any non-white skin colour that was perceived amongst native people resulted from their use of skin dyes and ointments. Reflecting the dominant view of the day, the Dutch scholar Jan Huygen (quoted by Vaughan, 1995, p.9) noted, in 1598, that when Native Americans 'come first into the world ... [they] are not so blacke but very white: the blacke yellowish colour is made upon them by a certaine oyntment.'[1]

Vaughan argues that the shift from 'white man to redskin' (as he puts it) occurred gradually over several centuries. Indications of this development may be found as early as the seventeenth century. By the mid-eighteenth century it had strengthened into the dominant regime of representation. As whiteness began to take a central place in the self-definition of Anglo-American settlers, native people were increasingly cast as 'tawny', 'brown' and 'red'. Reflecting this process, Nathaniel Rodgers (quoted by Vaughan, 1995, p.13), the editor of a 1764 edition of William Wood's natural history of New England, *New England's Prospect*, first published in 1634, felt moved to correct its author's view that native people 'are born fair'. '[T]his was one of the popular errors given into by our author,' Rodgers noted in his editorial introduction. This attempt to correct earlier errors reflected an emergent consensus that, as Benjamin Franklin had noted in his 1751 essay 'Observations Concerning the Increase of Mankind and the Peopling of the Earth', '[t]he number of purely white people in the world is proportionably very small.'[2]

Constructing racial whiteness

We may trace the attempt to construct an exclusionary and, eventually, highly racialised, interpretation of whiteness along a number of intersecting paths in European history. There are a variety of European national and regional articulations of white identity, the particularities of which lie beyond the scope of this chapter. My focus will be upon shared themes within the forms of white identity developed within Western Europe and the European settler societies. It is pertinent, first, to note that, as in China and the Middle East, there existed cultural traditions in ancient and medieval Europe that valued the colour white as a symbol of purity, religious devotion and nobility. The pale complexion attributed to aristocrats (according to pre-modern European legend, pale enough to see their veins, hence the expression 'blue blood') provided a physical marker of their noble descent. These traditions were woven with Christian representational tropes that privileged whiteness by associating it with chastity and godliness (Bastide, 1968). *The Holy Bible* (Douai version, 1609) is littered with such allusions, most famously within 'Solomon's Canticle of Canticles'. In the latter text, non-whiteness is seen as masking the true colour of a figure that is open to interpretation as humankind, the Church or the Holy Virgin: 'Do not consider me that I am brown, because the sun hath altered my colour' (verse 5). In verse 10, the latter speaker's 'beloved' (understood to be Christ) is described as 'white and ruddy, chosen out of thousands'.

It is interesting to note that the first usage of white as 'ethnic type (chiefly European or of European extraction)' that the Oxford English Dictionary cites, and which is not specifically applied to a non-European group, is the distinction made by an English cleric, C. Nesse, in 1680, between 'The White Line (the Posterity of Seth)' and 'the black line the Cursed brood of Cain'. Here whiteness and non-whiteness are being associated with distinct moral lineages, an association that was strengthened by interpretations of the post-Flood division of mankind. The widespread acceptance of the idea that, as expressed by the sixteenth-century chronicler George Best (quoted by Hannaford, 1996, p.166), Noah's sons, although '*all three being white*', founded the three branches of mankind, with Ham's progeny 'marked with a black badge to symbolize loathsomeness and banished to ... Africa where they lived as idolators' (Hannaford, 1996, pp.166–7), drew on a mixture of Jewish and Christian traditions (Jordan, 1968; Hannaford, 1996) to equate being heathen with being non-white. This association was reinforced by the intellectual fusion of the concept of 'Europe' with the older category of 'Christendom' that developed from the late medieval period onwards (Hay, 1957; Boer, 1995).

Thus a triple conflation of White = Europe = Christian arose that imparted moral, cultural and territorial content to whiteness. The broad constituency of this latter identity is suggestive of the transformation of the concept of race from a category denoting nobility, more specifically a noble line of descent, to the more socially inclusive idea of a people and/or nation. The earliest examples of this transition are to be found within Spanish colonial understandings of the peoples and social order of the New World. Themes of nobility, skin colour and Christianity, codified within the language of race in fifteenth-century Spain, were transmuted into a colonial discourse of white superiority and non-white inferiority. As Lewis (1995) explains,

> In the New World, defects of race, religion, and nationality were linked to skin color (not fully white) and ultimately to lineage (not fully Spanish/Christian) ... [T]he traditional Old World distinction between Spanish nobility and Spanish commoners gave way to distinctions among Spaniards, Indians, mixed-races, and blacks. The social hierarchy in New Spain was dominated by the white (Spanish) elite, the index of legitimacy. (pp.49–59)

Hannaford (1996; see also Guillaumin, 1995) evidences a similar social expansion of the race concept in Northwest Europe and its colonies from the end of the sixteenth century. His principal examples are English: Florio's (1968, first published 1611) assertion that race refers to 'royall blood' (cited by Hannaford, 1996, p.172); Holinshed's (1965, first published 1578) enlargement of the term to include 'Gentlemen' (cited by Hannaford, 1996, p.173); and Verstegan's (1605) deployment of the category to denote 'a people' (for example 'Englishmen are descended of the German race', cited by Hannaford, 1996, p.180). In English colonies, as in Spanish ones, the broadening constituency of 'race' facilitated its adoption as an integral part of ideologies of colonial expansion. As this implies, the ostensible social equivalence carried within the idea that, for example, as Jordan (1968 p.97) notes, within England's North American colonies, 'the terms *Christian, free, English,* and *white*' were 'employed indiscriminately as metonyms', needs to be understood within the context of the large-scale mobilisation (both physically and ideologically) of European-heritage peoples as agents of colonialism, imperialism and nationalism. That the earliest and most fully widespread employment of 'white' to refer to a European people, or European peoples, is to be found within colonial settings (see also Segal 1991) is symptomatic of the term's emergence as a key site in the forging of new mass political identities based on distinctions between coloniser and colonised, dominant and subject peoples.

The representational legacies of traditional notions of whiteness and race (nobility, piety and so on) were drawn upon, and into, new models of the world's population that asserted a hierarchy between races (with whites/Europeans at the top) and an equivalence within them. The category 'white European' became both a tautology and a group beyond compare. In Charles White's 1795 depiction, the 'white European', being most removed from the brute creation, may, on that account be considered as the most beautiful of the human race. No one will doubt his superiority in intellectual powers; and I believe it will be found that his capacity is naturally superior also to that of every other man. Where shall we find, unless in the European, that nobly arched head, containing such a quantity of brain? ... In what other quarter of the globe shall we find the blush that over spreads the soft features of the beautiful women of Europe, that emblem of modesty, of delicate feelings and sense? (cited by Fryer, 1984, pp. 168–9)

Non-European racial whites?

The expansion of European power legitimised and encouraged the development of racial science. However, the latter cannot be understood as merely the creation of the former. It was a semi-autonomous discourse, one capable of throwing up material that contradicted Europeans' attempt to claim an exclusive stake in whiteness. Many racial scientists drew upon cranial and linguistic investigations as well as, or instead of, skin colour to establish the boundaries of race. Sometimes these investigations were used to confirm that Europeans had sole claim to whiteness. Such work often sought legitimacy by reference to one of the earliest applications of scientific methodology to the interpretation of human difference, the second edition of Linnaeus's *General System of Nature* (published 1740; see Linnaeus, 1956). Using colour as his primary demarcator, Linnaeus proposed four basic human groups: 'Europaeus albus', 'Americanus rubescens', 'Asiaticus fuscus' and 'Africanus niger' (see also Blumenbach, 1997; first published 1776). Nevertheless, a much stronger current of scientific research supported the theory that Europeans were but one expression of a wider racial group (termed, sometimes interchangeably, Caucasian, Aryan and white), a group that included peoples from Asia and North Africa (for example, Smith, 1848; see also Gobineau, 1915). This tradition established itself as the more scholarly expression of racial science. Thus we find it propagated in nineteenth- and twentieth-century student texts, for example in the English schoolbook *A Geography of Africa* (Lyde, 1914, first published 1899). 'The non-European population' of Africa, the book explains, 'belongs mainly to one of two races, the White and the Black.' Amongst the whites of Africa are included 'Arabs and Abyssinians ... Berbers and Tuaregs, Masi and Somalis' (p.2).

However, the inclusivity of such opinions is deceptive. Their apparent extension of whiteness to non-Europeans was consistently undermined both by more powerful, and more popular, discourses affirming European racial supremacy and by its own proponents' tendency to find Europeans to be the most authentic and best exemplars of the white race (see Jacobson, 1998; Lopez, 1996). For a typical example of this latter argument we may turn to Sir Arthur Keith's (1922, p.xv), introduction to the popular anthropological series *Peoples of all Nations*. He notes that, although the Caucasian race 'extends from Southern India to Scandinavia. At the European end of this line we find the cradle-land of the blond man; at its Indian end we find peoples showing distinct Australoid and Negroid traits.' The weakness of inclusive views of whiteness was even more clearly signalled in the 1920s by the reaction of the American Supreme Court to an Indian migrant's application for naturalisation as a white citizen:

> the blond Scandinavian and the brown Hindu have a common ancestor in the dim reaches of antiquity, but the average man knows perfectly well that there are unmistakable and profound differences between them today. (cited by Sacks, 1994, p.81)

Of course the assertion that the 'white race' is not coterminous with Europeans continues to be articulated today. However, within Europe and North America, it has been reduced from being a significant rival to more exclusionary views, to the status of a technicality little favoured outside certain immigration bureaucracies and traditional anthroplogy.[3] Thus United States Department of Immigration's Directive 15, which defines North African and Middle Eastern peoples as white, is increasingly treated as an oddity. Mostafa Hefney, an Egyptian immigrant who recently legally contested the directive, could find no other explanation for its existence than the fact it 'provides Whites with legal ground to claim Egypt as a White civilisation' (quoted by Morsy, 1994, p.176).

Non-European identities in transition

The central imperative behind the West's racialisation of the world was to legitimise Europeans as part of a superior race. The ideology of race, with its concomitant divisions, was disseminated across the globe, making profound inroads into non-Europeans' sense of self and other. Within some societies this process appears to have overlain, or merely marginalised, more traditional definitions of identity. In the Middle East, for example, Western notions of race and, subsequently, Social Darwinism, did not completely erase the autonomous tradition of referring to Middle Easterners as white (for an account of the reception of Darwinism in the Middle East, see Ziadet, 1986). 'Africa is divided into Black and White', notes Fanon (1967, p.129) in *The Wretched of the Earth*. This 'commonest racial feeling', he continues, is based on the conviction 'that White Africa ... is Mediterranean, that she is a continuation of Europe' (pp.129–30). '[T]his Arab nation,' records another, somewhat more recent, commentator on Arab politics, 'comprises the black and the white within the limits of this vast land' (Azzam, 1976, p.170; see also Brown, 1968). However, such affirmations have an increasingly residual, dated ring to them. Modern Western and Middle Eastern and North African notions of identity have effectively marginalised – and, sometimes, actively resisted – this particular usage of the term 'white'. Thus, within twentieth-century international politics, Middle Eastern and North African nations have consistently aligned themselves as, and with, non-white and non-Western powers (Haim, 1976; Hourani, 1970). The development of 'Arab nationalism', with its concomitant sense of ethnic and racial particularity (see, for example, Rabbath's essay 'The common origin of the Arabs', 1976), has encouraged a resistance to cultural, political or racial identification with Europe. For some Middle Eastern commentators, this process necessitates an explicit rejection of claims to whiteness. The Egyptian scholar Soheir Morsy (1994, p.180) offers a personal and generational explanation of this transition: 'In my mother's generation, the desirable attributes for a bride among the upper classes were a fair complexion and the ability to speak French.' In contrast, Morsy herself, influenced by pan-Africanism and pan-Arabism, locates whiteness firmly within Europe and North America.

The impact of the Western claim to whiteness appears, at first sight, to have made an even more profound impact in China than in the Middle East. 'White' was abandoned as a self-definition by elite Chinese and the term 'yellow' widely adopted. Whiteness retained its positive symbolism but was increasingly seen as an attribute solely possessed or best exemplified by Europeans (Wilson, 1984a; 1984b; Bernstein *et al.*, 1981). By the late nineteenth century, racist, Social Darwinian interpretations of the struggle for survival and supremacy between the world's races were being offered by a new generation of Chinese intellectuals. The yellows, it was argued, could share the intellectual and leadership capabilities of the whites. Dikötter (1992, p.81) cites the late-nineteenth-century writer Tang Caichang's interpretation of the categories of Western racial science, 'Yellow and white are wise, red and black are stupid; yellow and white are rulers; red and black are slaves; yellow and white are united, red and black are scattered.'

However, the Chinese designation of themselves as yellow evidences more than the simple absorption and regurgitation of Western racial science. As Dikötter emphasises, the Chinese viewed Western science through the prism of existing, and mutable, representational traditions. In respect to the changing role of white identities, it is important to note, first, that, in China, the term 'yellow' may include certain straw or blond forms of whiteness. Moreover, yellow was the imperial colour and highly positively connoted, symbolising fertility, fame and progress (Chevalier and Gheerbrant, 1996). In addition, as we have seen, although white symbolised certain esteemed qualities, it also signified death. Thus the adoption of 'yellow' and the erasure of China's 'white past' was, in some measure, enabled by that society's specific traditions of chromatic symbolism. A fascinating reflection on this process is contained in a letter written by a Chinese scholar during the last years of the Manchu dynasty:

> Of the five colors, yellow is the color of the soil, and the soil is the core of the universe. Westerners identify Chinese as a yellow race. This implies that from the beginning, when heaven and earth were created, the Chinese were given the central place. When Westerners laugh at Chinese egotism, why can we not explain it by this reasoning? (quoted by Isaacs, 1968, p.92)

The instabilities of racial whiteness

Modern European white identity is historically unique. People in other societies may be seen to have valued whiteness and to have employed the concept to define, at least in part, who and what they were. *But they did not treat being white as a natural category nor did they invest so much of their sense of identity within it.* Europeans racialised, which is to say naturalised, the concept of whiteness, and entrusted it with the essence of their community. Europeans turned whiteness into a fetish object, *a talisman of the natural* whose power appeared to enable them to impose their will on the world.

This development necessitated the denial of other forms of white identity. One startling exemplification of the success of this process may be found in the pages of the anthropologist Marsh's *White Indians of Darian*, published in 1934 (see Taussig, 1993 for discussion). Marsh had heard stories of the existence of 'white Indians' in the Darian area of Panama and, incredulous but fascinated, set out to see for himself. He can scarcely believe his eyes when he first catches sight of such an individual – an Indian who, says Marsh, *looked like a European*.

> He certainly made a strange appearance among his dark-skinned countrymen. His hair was light golden yellow. His skin was as white as a Swede's. ... I looked at him in amazement. Here was my white Indian at last ... the scientists would have a grand time explaining him. (Marsh, 1934, p.199)

Thus the very concept of a non-white European is reduced to the level of an anthropological freak, a bizarre tale guaranteed to drop jaws all over Europe and North America.[4]

Claims of European white difference and superiority informed almost every aspect of the social and political life of Europe and the territories it dominated. The obsessive, excessive nature of Europeans' discursive employment of whiteness is one of its most intriguing, and increasingly discussed, features. In *Colonial Desire*, Young (1995) provides a number of largely nineteenth-century examples of the extraordinary breadth of Europeans' investment in white identity. Much of his material refers to the way the racialisation of whiteness necessitated a set of prohibitions and 'moral panics' around inter-racial sex. However, Young also discusses the demands racial whiteness made upon the nature of war and the writing of history. On the former point he (p.7) quotes the German anthropologist Theodor Waitz's claim of 1859 that 'All wars of extermination, whenever the lower species are in the way of the white man, are not only excusable, but fully justifiable.' Dreams of genocide also seem to structure the racial scientist Josiah Nott's (1844, p.16) historical studies, particularly his conclusion that the 'adulteration of blood is the reason why Egypt and the Barbary States never can rise again, until the present races are exterminated, and the Caucasian substituted.'

European racial whiteness is an extraordinarily ambitious social project. It makes enormous demands, both on 'other races' and upon its European progenitors. The latter are required to be the epitome of civilisation, of purity, of morality and of virulent strength. The central irony of this process is that this set of expectations is placed on societies riven with, and based on, the existence of hierarchies that deny these very qualities to most of their population. Whiteness, both as the ultimate symbol of superiority and as the legitimising authority and mobilising ideology for national imperial and colonial enterprises, must simultaneously be made available to all Europeans and denied to those deemed unfit or unwilling to carry its burden. Thus the excessive nature of the European construction of whiteness, its exclusionary zeal, brings about its own impossibility: most whites are unworthy of whiteness.

This contradiction is something historically distinct within European ideologies of human difference. The concepts of race and whiteness are traceable to religious and aristocratic traditions, traditions that denied their application to most Europeans. Each category has a long history of ethnic, class and gender exclusions. However, both the rise of political forms which required the mass mobilisation of European-heritage peoples as colonial and national agents, and the intimately related inten-sification and reification of the positive connotations of whiteness, may be seen to have placed whiteness on a collision course with discourses of difference that subverted its claims of social equivalence. In order to explain this proposition more fully, I turn now to three examples of social hierarchy – ethnicity, class and gender – that cut across and disturb racial whiteness. I shall address the first two of these examples, ethnicity and class, only briefly before examining, in more detail, the obsessive and continuously failing assignment of whiteness to women that occurs in Henry Champly's (1936) *White Women, Coloured Men*.

The inclusion of different European heritage class and ethnic groups within whiteness mirrors, at least in part, these groups' inclusion within particular colonial and national projects. The examples of the Irish and Italians in the USA and Britain and the Victorian urban working class may be used to illustrate this point. The marginalisation of certain Euro-pean ethnic groups from whiteness has been documented in some detail in recent research on nineteenth- and early twentieth-century American immigration and labour relations regimes. These studies have shown how whiteness was initially denied to those European immigrant groups (such as the Irish and Italians) who were socially and economically excluded from the Anglo-American elite. Thus, for example, in Ignatiev's (1995) *How the Irish Became White,* a transition from exclusion to inclusion into both 'Americanness' and whiteness is traced. Ignatiev argues that this process may be explained by reference to Irish immigrants' increas-ing economic power and ability to be accepted into America's racialised labour market 'on the side' of 'the whites'. Roediger has charted a parallel and overlapping set of histories in *The Wages of Whiteness* (1992) and *Towards the Abolition of Whiteness* (1994). For Ignatiev and Roediger, this history evidences the fact that whiteness – as a central signifier of status and power – cannot remain undisputed in a racialised economy. Thus they show how the domain of European racial whiteness can expand and contract, how it is literally and symbolically fought for amongst Europeans.

This history is not confined to the USA. Within Europe there also exists a wide variety of histories of ethnic exclusion from whiteness. In Britain, the otherness of the Irish, their status as colonial subjects rather than agents, has been marked both by their Victorian scientific identification as a lower race and by their persistent cultural representation as non-civilised and primitive (Innes, 1994; Curtis, 1997; Cohen, 1988). Another, richly ironic, example of a minority group being excluded both from whiteness and from a particular national/colonial identity is the tendency amongst contemporary Russians to revile the peoples of the Caucasus as 'blacks' (ironic because according to one of the earliest theories of racial science, the Caucus Mountains are the ancestral homeland of the white race, hence the term 'Caucasians'). In an account of the assault of 200 ex-paratroopers on ethnic 'Caucasians' in Moscow in 1994, Meek (1994) quotes one policeman's response when summoned for help, 'When these blacks rape your daughters, you'll be complaining. Let the guys sort them out.'

As I explain in Chapter 2, the inherently exclusionary nature of racial whiteness may also be seen at work within Victorian class relations. Perhaps the best-known expression of the assumption that upper-class Britons were literally more white than working-class Britons is the remark attributed to Lord Milner on seeing English soldiers washing, 'I never knew the working classes had such white skins' (attributed by Cohen, 1997, p.256). Given the class-bound nature of the allocation of whiteness it is not surprising that the mere act of travel from one class-identified area of the city to another was often expressed through colonial metaphors. 'As there is a darkest Africa, is there not also a darkest England?' asked William Booth in *In Darkest England and the Way Out* (1976, p.145, first published 1890).

The development of racial whiteness in societies structured upon ethnic and class division inevitably produces contradiction. Because of its function as the symbol of European supremacy, a fetishised and excessive ideal, any undifferentiated application of racial whiteness to, for example, all Britons, would provide a profound challenge to class and ethnic hierarchy within that society. In other words, if working classes and outcast minorities were accorded the same level of whiteness as Lord Milner, the *rationale* for their lowly social location – i.e. their depravity, primitiveness, stupidity, darkness – would be placed in jeopardy.

We have seen that the construction of racial whiteness has had profound social consequences. It cannot exist alongside other white identities. In the present century, the ideals, norms and distinctions associated with racial whiteness have affected every society, touched every life. Yet the excessive idealisation inherent within racial whiteness inevitably creates contradiction and potential crisis. My last instance of this process provides one of the most telling examples of the fragility of racial whiteness.

The gendered nature of whiteness has been analysed from a variety of perspectives (for example, Frankenberg's (1993) interview-based, experiential studies, Ware's historical sociology (1992) and Dyer's (1997) analysis of racial imagery). However, nearly all commentators concur that women have been assigned specific functions within the racial ideal, namely of chastity, purity and as biological reproducers. Since racial whiteness is ideologically rooted in naturalism, the central expectations and assumptions made about white women are gathered around this latter, physical, role. As expressed in 1906 by Gemmell (quoted by Ware,

1992, p.35) in the *Journal of Obstetrics and Gynaecology of the British Empire*, the 'supreme purpose' of women was 'the procreation and preservation of the race'. Ware's work provides the most useful material for historically informed approaches to this process (see also Beezer, 1993; Bland, 1982). Drawing material from the late nineteenth century, she notes that

> women were seen as the 'conduits of the essence of the race'. They not only symbolised the guardians of the race in their reproductive capacity, but they also provided – as long as they were of the right class and breeding – a guarantee that British morals and principles were adhered to in the settler community, as well as being transmitted to the next generation. (pp.37–8)

The positioning of women as the carriers of the white race, combined with more traditional ideas concerning females' chastity and civilising influence, meant that the burden of racial whiteness was visited more intensely upon females than males.[5] This responsibility was a form of 'white privilege' of a very particular kind. For, whilst placing women 'on a pedestal', it demanded their submission to a wide range of social controls designed to protect them from corrupting influences. This process was, in turn, identified and resisted by some women, a subversion that resulted in their identification as a threat to the white race (Ware, 1992). As this brief overview suggests, the gender politics of racial whiteness provides one of the most complex examples of its crisis-prone ideological constitution.

In order to tease out and exemplify the gendering of racial whiteness, I will turn to Champley's (1936) *White Women, Coloured Men*. Champley was a French newspaper editor who, during the inter-war period, wrote a number of mass circulation books on the theme of race relations. The particular text I am concerned with here (a translation from the French original) is a somewhat salacious yet also very serious account of his observations, made in various parts of the world, on the sexual attraction between white women and non-white men. The supposed threat of 'colour' against 'the white race' was already well established as a site of crisis and controversy by the time Champley's book was published. In *The Conflict of Colour*, published in 1910, Putnam Weale had warned that a 'struggle has begun between the white man and all other men of the world' (p.98). Japan's military defeat of Russia in 1905 combined with the sense that white solidarity and power were being undermined by inter-European rivalry (of which the First World War was the most striking example) seemed to many commentators to herald what Leo Chiozza Money in *The Peril of the White* (1925) called 'race suicide' (see Füredi, 1998 for discussion). The 'authority of the white man', wrote Matthews in *The Clash of Colour* (1924, p.30) 'has ended'. In the United States, Grant Madison's (1916) *Passing of the Great Race* and Lothrop Stoddard's tract *The Rising Tide of Color Against White World Supremacy*, which appeared in 1920, provided similarly doom-laden warnings. Nevertheless, for my purposes what is of interest in Champley's work is not its representativeness of 'white attitudes' but, rather, the way it explicitly engages with and exhibits the logical implications of the traditions of racial whiteness.

In other words, the book is useful because of its exhaustive 'working through' of the burdens and claims of racial whiteness.

Champley firmly locates racial whiteness in the body of white women. Moreover, he does not shy away from the implication of this positioning, that whiteness cannot and should not reside fully in the male. On seeing a group of young adults playing on a beach he notes that 'The girls were the paler, *just because they were female*; the boys were warmer-coloured, as males should be' (p.24; the lighter colouring of women in Western racial imagery of whites is explored in Richard Dyer's excellent study *White*, 1997). Thus the fetishisation of whiteness is organised around the figure of the white woman. And, logically enough, all the 'other races' of the world desire her. Champley's basic argument, what he calls his 'formula', is as follows:

> The Coloured peoples have discovered the White woman – as a marvel; as a wonder from the physical, the artistic, the social and even the religious points of view; as an idol worthy of being desired above all else. ... *Beware, White race! The Coloured races have discovered your supreme treasure, the White woman!* (pp.6–7)

Champley provides numerous examples of the 'coloured man's' rapacious need for white women. Drawing on examples from China, he notes that for white men to lose 'the struggle between Yellow men and White men for the conquest of White women' (p.242) is to lose possession of whiteness, hence, world dominance and, by implication, identity itself.

Champley also asserts that, because they embody the essence of whiteness, white women have power over the fate both of the white race in general and of white men in particular. Indeed, *White Women, Coloured Men* is structured around a set of barely concealed panics concerning female moral and political autonomy. The author's basic conundrum is, to paraphrase, 'Will they save us or abandon us?'. His jotted 'dedication', a poetic plea made on the cover of the book's manuscript (and transcribed in its published form), reads as follows:

> Women of Great Britain or Germany, Italy or Norway, Australia or America, women of *my race*:
> *White women*, you who are the most highly charged with sex-appeal, the most desirable in the world –
> *To-morrow the world will be fighting for you ...*
> Will there, indeed, be any White women left in a hundred years hence? But you are the most understanding of women. What if your feminine genius should dominate the situation, preserve the race, and solve this almost insoluble imbroglio?
> Then, for all of you, for all those who love you, for your countries, for your descendants, it would mean salvation. (p.253)

Such exaggerated, almost mocking, male prostration before the will of women is a recognisable and traditional trope of male-dominated societies. However, amongst all the patronising bluff, there is also a sense of genuine unease and bewilderment. For Champley's travels have convinced him that white women are deserting the cause, they are leaving white men for 'coloured' men. This racial treason, he contends, has two principal roots, female expressive freedom and racial impurity. The latter argument is developed somewhat half-heartedly. Even Champley cannot, it seems, quite convince himself that only women who 'belong to less pure racial categories' are 'attracted by the exotic charm of Coloured men' (p.37). He is more assertive about the racially subversive nature of feminism. In a chapter titled 'The white women: our greatest sin', he argues that women's autonomous expression of sexuality must be curtailed. Such *'sexual disturbers of the peace'* (p.250) must be challenged and returned to a condition of 'heroic humility' (p.310).

One of the central ironies of racial whiteness is that, although it developed as an ideology of domination, its heaviest burden was carried by a subjugated group, i.e. women. Yet, as Champley's writings indicate, even its most ardent propagandists have doubted the sustainability of racial whiteness. Because of its excessive and fetishistic nature, it is constantly endangered by groups (for example, women, the Irish and the working class) who fail to live up to, or must be denied, its extraordinary demands.

Whiteness has been the central signifier of Europeans' superiority: it cannot be shared. Yet it is this very exclusionary zeal, with its concomitant excessive idealisation, that provides this unique form of collective identity with its greatest challenge.

Conclusions

The development of racial whiteness demanded the disappearance of non-European white identities. However, this is not to argue that the latter set of identities was necessarily actively obliterated or suddenly replaced. Within China and the Middle East, European ideas and power were engaged and interpreted through both traditional and modern discourses of representation. Moreover, non-European white identities still exist, albeit at a relatively residual level (for example, in the Middle East). Some American commentators have also recently suggested that whiteness may, over time, be expanded in the United States to include certain East Asian American groups, such as Japanese Americans (Hacker 1992; see also Chapter 3).

However, despite these provisos, it seems incontrovertible that the main event in the history of white identity has been its reinvention by Europeans over the last 250 years. Indeed, European military and socioeconomic power has proved capable not only of claiming whiteness as a uniquely European attribute, and of getting other people to believe this, but also of erasing the fact that white identities *ever had a history outside Europe*.

This chapter has tried to expose this power to make history and geography disappear. To this end, it has inevitably had to transgress the narrow historical and geographical specialisations that tend to characterise racial and ethnic studies. The particularity of European forms of racial whiteness cannot come into view if we contemplate only the modern era and only Europe and America. Only within a longer and wider context can the process of whites becoming non-whites and Europeans becoming the only whites be brought to visibility.

1. A similar conception about Africans also survived well into the eighteenth century. In his multi-volume *Histoire naturelle générale et particulière* (first published 1748–1804), Georges-Louis Leclerc, Comte de Buffon, commented upon the extent of this belief and noted 'that the air is necessary to produce the blackness of Negroes. Their children are born white, or rather red, like those of other men. But two or three days after birth their colour changes to a yellowish tawney, which grows gradually darker till the seventh or eighth day, when they are totally black' (Buffon, 1997, p.23). White births among Africans also exercised the imagination of a contemporary of Buffon's, Pierre-Louis de Maupertuis (cited by Polia-kov 1974, p.164), who, however, accounted them as incidences of albinism, an interpretation that led him to suggest 'that white is the primitive colour of mankind while black is only a variation which has become hereditary in the course of centuries but which has not entirely effaced the white colour, the latter always tending to reappear'.

2. Indeed, reflecting what was to become a characteristic tendency within US race dis-course, Franklin attempted to differentiate the categories 'European' and white', concluding that the fomer was a rare and precious attribute, and one largely owned by the Saxons and the English (see Jacobson, 1998 for a discussion of this debate).

> All Africa is black or tawny; Asia chiefly tawny; America (exclusive of the newcomers wholly so). And in Europe, the Spaniards, Italians, French, Russians, and Swedes are gen-erally of what we call a swarthy complexion; as are the Germans also, the Saxons only excepted, who, with the English, make the principal body of white people on the face of the earth. I could wish their numbers be increased. (Franklin, cited by Jacobson, 1998, p.40)

3. Such traditionalists can also be found in sociology, such as the author of the book whose title has been mutated in the service of the present chapter. In *Who is Black?* Davis (1991) uses an arsenal of confusing 'scientific' categories to expose European and African Ameri-cans' 'mistaken' identification of Arabs and Indians as non-white: 'Most people from India are certainly dark, but they are so-called Hindu caucasoids'; further 'Most Arabs are Medi-terranean whites, although many in Iraq are descended from Hindu caucasoids … as are other Arabs and many Iranians' (p.162).

4. Some 'white Indians' were taken to Washington and examined by scientists. They were declared not to be 'white' but 'partially' albino. Taussig (1993, p.274) cites the American geneticist R.G. Harris's explanation, published in 1926, that 'The White Indians obviously express a form of albinism which has been termed imperfect or partial albinism by Geoffroy Saint Hilaire, Pearson and others. These terms signify that either the skin, hair, or eyes, any two or all three may fail to express the full albinotic condition, but that one or more are, par-tially at least, relatively free from pigment.' For a more recent version of 'white Indians' as anthropological freaks and tourist attractions see the *Peru Handbook*'s (Murphy, 1997) account of how tourists can include a trip to 'the village of the whites' in their Peruvian hol-iday. The 'incongruous' inhabitants of this remote community, the guidebook notes, 'could easily have just stepped off the five o'clock flight from Stockholm' (p.193) and are claimed to be descended from Vikings.

5. The association between whiteness and femininity is not confined to Europe. Thus, for example, Isaacs (1968) notes the existence of this link within ancient China. Moreover, one of the fullest treatments of the topic emerged from a Japanese archival and interview-based study conducted by Wagatsuma (1968). Wagatsuma cites Japanese men living in the United States as noting 'Whiteness is a symbol of woman, distinguishing them from men', 'One's mother-image is white' (pp.139–40).

CHAPTER 2

How the British working class became white

I never knew the working classes had such white skins. (Lord Milner on seeing English soldiers washing during the Battle of the Somme, attributed by Cohen, 1997, p.256)

Introduction

In the nineteenth century, the notion that all Britons were white was asserted with considerably more force and conviction outside Britain than within it. From a colonial distance it was a commonplace to accord a white, and hence elite, identity to every inhabitant of the United Kingdom. On home soil, however, the assertion that everyone was equally white was more problematic. The British working class, for example, were marginal to the symbolic formation of whiteness and sometimes, as Lord Milner's remark implies, actively excluded from it.

In mid–late twentieth-century Britain, as in other European societies, the meaning and boundaries of whiteness have changed beyond recognition. Today politicians and other commentators feel apparently unconstrained about employing the concept to depict the majority of 'ordinary', working-class Britons.

This chapter seeks to contribute to the explanation of this transition. Thus I wish to investigate why and how a group once marginal to whiteness could later come actively to employ this identity as if it was significant – or, indeed, central – to their own sense of self, nation and community. My analytical focus will be upon how shifts within the connotative repertoire of, and relationship between, class and whiteness have enabled, and been enabled by, different forms of capitalist socioeconomic organisation. The perspective I will be offering on these themes is an admittedly partial one. The ethnic and gendered nature of whiteness, which I discuss elsewhere in this book, is only briefly addressed. My intention in this chapter is not to provide a panoptic survey of the diversity of white identities but something more limited: to offer a number of insights, based on the analysis of the changing nature of capitalism and the representation of class identity, into why and how the working class with a 'European ethnic heritage' adopted and adapted whiteness in one particular country.

Historical studies of the relationship between class and white identity have recently enjoyed something of a boom. However, as I noted earlier, nearly all the new scholarship in this area has emanated from, and been concerned with, the USA. Despite the value of this work the racial history of the latter country cannot be used as a template for other societies. White identity was incorporated into US politics and economics comparatively early, from the late seventeenth century. US workers were assertively white many generations before the British working class may be found politically organising around 'their' whiteness (a phenomenon which did not occur to any significant degree until the 1950s). Moreover, relative to Victorian Britain, white identity in the USA was rarely explicitly class-exclusive. Rather, the language of ethnicity came to submerge the language of class. Thus, for example, Irish and Italian immigrants were not initially construed as indisputably white, but all Anglo-Americans (and, from the late nineteenth century, all European-heritage Americans), of whatever class, were.

Thus 'How the British working class became white' is a different story from Noel Ignatiev's (1995) account of *How the Irish Became White* in the USA. The development of British white identity cannot be understood as representing a belated re-enactment of the American experience. Another seductively simple interpretative error also requires our attention. For it is, perhaps, just as tempting to imagine that the single catalyst for the assertion of working-class whiteness in Britain was non-white immigration into that country: white identity arrived in working-class politics when, and because, 'people of colour' arrived in Britain. The argument that working-class racism developed out of a perceived competition between 'white residents' and 'non-white immigrants' for scare resources, such as housing and jobs, has all too often implied the same conclusion (for example, Phizacklea and Miles, 1979). It is presupposed that the 'white working class' is an unproblematic category: that the white working class always were white, always knew they were white, and merely required the presence of non-white communities/competition in order to start mobilising around this 'fact'. Thus any sense of historical transformation in white identity is lost, and any discussion of how and why the working class were able to start drawing on a form of social symbolism from which they had once been marginalised, effectively prevented.

Chapter 2 shows that, although non-white immigration provided a catalyst for white working-class *deployment* of white identity, the fact that this identity was *available* to, and adapted by, European-heritage working-class Britons, demands an analysis of the changing symbolic constitution of racialised capitalism. The chapter has two sections. Both draw on the accounts of those who, implicitly or explicitly, claim possession of, or a clear stake in, white identity. In the first part, I look at the symbolic production of whiteness in nineteenth-century Britain. Drawing on published sources written for and by middle- and upper-class

Victorians, I suggest that metaphorical and literal depictions of racial whiteness were employed as a new paradigm of class hierarchy, a paradigm imported into Britain from colonial and settler societies. Whiteness was fetishised and idealised as an 'extraordinary', almost superhuman identity; an identity developed, in the main, for and by the bourgeoisie. The Victorian working class, most particularly the urban and immigrant working class, were positioned as marginal to this construction. In my second section, I discuss the transformation of this tradition from the late nineteenth century. I suggest that whiteness became available to the working class because of changes within the socioeconomic and symbolic structuring of British capitalism (more specifically, I discuss imperialism and the rise of the welfare state). The aggressively defensive articulations of working-class whiteness examined in this section bear the imprint of these changes, more specifically a shift in emphasis from whiteness as a bourgeois identity, connoting extraordinary qualities, to whiteness as a popularist identity connoting superiority but also ordinariness, nation and community. Appended to this discussion is a commentary on how recent reformations of welfare capitalism may effect its racialisation.

Working-class marginalisation from white racial identity in Victorian Britain

The racial categories 'white' and 'European', whilst first intellectually delineated in Europe, had their earliest and profoundest socioeconomic articulation outside Europe. Thus, for example, we find these racial categories being employed, as early as the sixteenth century, to legally and economically structure colonial and settler societies in South and Central America and, from the seventeenth century, to perform the same task in North America. In these non-European settings, ascriptions of whiteness translated directly into socioeconomic status.

In marked contrast, even by the early nineteenth century, discussion of racial whiteness in Britain was focused outwards, being dominated by references to colonial and settler societies and, more generally, Europe's role in the world. It is within this debate that we first find the character and nature of whiteness being explicitly fleshed out in Europe, its scientific delineation amplified into a moral and aesthetic vision.

In the course of the nineteenth century, racial and class meaning became increasingly blurred into a mutually constitutive set of associations and images. The influence of class as a paradigm for racial inequality has been asserted by a number of commentators (Lorimer, 1978; Malik, 1996; Benedict, 1940). However, just as apparent by mid-century was a flow of ideas in the opposite direction. In other words, racial categories associated with the colonial project and settler societies were having a clear impact upon the British elite's metaphors of class domination and hierarchy. Thus, for example, images of colonial voyaging and the conquest of 'dark' regions were drawn upon to narrate and justify bourgeois incursions into working-class environments (for example, Booth,

1976). Similarly, the influence of the colour-coded language of hierarchy used in settler and colonial societies provided an alluring formula to articulate Britian's own social divisions. The influence of the binary 'black and white' model of social difference associated with the USA appears to have been particularly strong.[1] It is pertinent to note, in this regard, how influential novels and journalistic accounts concerning the 'colour divide' in US society were in Victorian Britain. Stowe's (1852) *Uncle Tom's Cabin; or, Life Among the Lowly*, which drew comparisons between US slaves and the English working class, was the best-selling novel in nineteenth-century Britain.[2] Another influential cultural import that offered a dualistic vision of society was 'black and white minstrel' shows. They first appeared on the English stage in 1836, and Lorimer (1978, p.86) notes that by the 1860s fifteen permanent minstrel companies had been formed and a 'craze for the minstrels had taken hold'. Lorimer (p.11) suggests that 'the Victorian (and Edwardian) looked upon the Negro as the photographic negative of the Anglo-Saxon'; that 'they seemed to get a clearer perception of their own supposed racial uniqueness from the inverted image of the black man'. This contention is supported by Pickering's studies of 'nigger minstrelsy' in nineteenth-century Britain. The '"nigger" mask', Pickering (1986, p.90) claims, 'symbolised all that was the opposite of the bourgeois conception of the white English person'.

As we shall see, by late mid-century, allusions to, as well as metaphors and literal depictions of, racial and colour difference had became integrated into the way working-class distinctiveness was understood by the middle and upper classes. However, the reason this happened is more complex than simply foreign and colonial 'influence'. We must also consider how this influence was mediated and enabled by Victorian capitalism.

As in contemporary Britain, the class order of Victorian Britain was legitimised through imputed class attributes and values. However, in the nineteenth century, this order had adapted only relatively superficially to the challenges of organised labour and the complexities of mass consumerism. As a consequence Victorian capitalism largely did without the kind of state interventions in social welfare that characterise its contemporary form. This apparently laissez-faire phase of capitalist development both enabled, and was enabled by, the formation of ideologies that naturalised and reified social difference. More specifically, the inequalities of wealth and power between the working class and the bourgeoisie were legitimised and justified as reflecting a natural order of talent: the inevitable side effect of the existence of economic growth (as explained by Adam Smith, 1910; first published 1776) and superior and inferior types of people (see Faber, 1971; Razzell, 1973). The bourgeoisie's ability to situate themselves as the most authentic exemplars of racial and national ideals, such as the Anglo-Saxon, the Briton and the white race, were central to this naturalisation process and, hence, integral to the reproduction of liberal capitalism.

However, before exemplifying how this process acted to marginalise and exclude the British working class, it is necessary to note the existence of another, more indigenous, and pre-modern, tradition of white identity. It would be misleading to suggest that white identities (but not, I would emphasise, white *racial* identities) did not exist before the development of racial science or, indeed, colonialism. The two most significant examples of this pre-modern legacy are the association of whiteness with religious devotion and purity (see Bastide 1968) and the related association of whiteness with high social status (Chevalier and Gheerbrant 1996). The latter association, which is somewhat less well known than the former, links together aristocracy, leading a leisured and sheltered (both literally and metaphorically) life, and possessing pale skin. Indeed, the meto-nymic encapsulation of the nobility's superiority, namely the expression 'blue blood', is derived from the myth that aristocrats have skin so trans-parent, so white, that their arterial blood may be seen. In the context of an ascendant bourgeois class, it is interesting to observe that aristocratic pal-lor came to be used as a metaphor for the decadence of the upper class in the late nineteenth century and twentieth century. From the ash-white Count Dracula (Stoker, 1897), with his thirst for rejuvenation, to the pale and dolorous inhabitants of Mervyn Peake's (1985) *Gormenghast*, a discourse emerged which depicted aristocratic social redundancy as physically incarnated in the faltering and feeble nature of their whiteness. These were white people who had lost their vitality; white people with-out blood, dusty and crumbling.

By contrast, bourgeois whiteness came to provide the category's hege-monic meaning, symbolising a wide and highly aspirational set of social ideals; a synonym for a healthy and vigorous civilisation and also, by extension, for ambitious and expansionist capitalism. Indeed, to indicate its excessive, fetishistic, character we might term this racial formation 'hyper-whiteness'. By the mid–late century, this excessive idealisation may be found contributing to the ideological legitimisation and forma-tion of almost every aspect of the social and political life of Britain and the territories it dominated.

Victorian racial whiteness was an extraordinarily ambitious social project. It made enormous demands upon its progenitors. The central irony of this process is that these expectations were placed on societies riven with, and based on, the existence of harsh and highly exclusionary social hierarchies. Whiteness, as the phenotype of civilisation, must simultaneously be made available to all Europeans within the colonial imagination but denied to those deemed unfit or unwilling to carry its burden within Europe itself. Thus the British working class was 'white' in colonial settings (including imaginative transposition of colonial settings to Britain; for example, Robert Sherard's (1897) *White Slaves of England*) but something less than, or other to, white in the context of Britain's inter-nal social hierarchy. In the latter context, the excessive nature of the bour-geois construction of whiteness, its exclusionary zeal, brings about its own impossibility: most whites, at least within Britain, were unworthy of whiteness.

If we accept that bourgeois whiteness performed a normative role in definitions of whiteness in the nineteenth century, it follows that the greater the cultural or economic distance that existed between a non-bourgeois group and this class, the less authentically white that group was likely to be considered to be. Unsurprisingly, those elements of the working class understood to be the most deferential to, and aspiring towards, a bourgeois identity tended to be represented by the Victorian commentators addressed in this section as less 'foreign' and, metaphorically at least, less 'dark' than others of their class. Thus the industrious 'labour aristocracy' were differentiated from the criminally inclined and work-shy 'residuum' (on the former see Gray, 1981; on the latter Stedman Jones, 1971). It was the 'residuum' (a term that is comparable to the contemporary category 'underclass') that formed the focus of middle class fears and hostility. However, the boundary between the different economic sectors of the working class was both broad and permeable, particularly during periods of social unrest. Indeed, as we shall see, the construction of the 'respectable' working class as aspiring towards bourgeois values was persistently undercut by a suspicion of all the 'urban masses' as harbingers of an uncivilised and alien way of life.

The refusal of authentic racial whiteness to the working class may be witnessed operating in two main ways: through imaginatively aligning them with non-whites, and by asserting that they are literally racially distinct from the middle and upper classes. I shall introduce examples of both representational attitudes, starting with the first. I shall also be touching upon the fact that the class rationale of these processes was often mingled with, or overlaid by, distinctions between the urban and rural working class as well as between Irish and English.

One of the most widely used metaphors applied to bourgeois contact with the working class was drawn directly from colonial practice, namely exploration. The writers of such explorations cast themselves as gallant yet concerned denizens of a more civilised land. They embodied the characteristics of hyper-whiteness: figures of tremendous energy and moral rectitude who had little in common with the 'heathens and savages in the heart of our capital' (Booth, quoted by Keating, 1976, p.19). Thus the mere act of travel from one class-identified area of the city to another was frequently expressed in the racialised language of imperial voyaging and conquest. 'As there is a darkest Africa, is there not also a darkest England?' asked William Booth in *Darkest England and the Way Out* (1976, p.145; first published 1890). As with the other accounts cited here, Booth employs 'black Africa' as his central simile of 'dark England'. Within 'a stone's throw of our cathedrals and palaces', he continues, exist

> similar horrors to those Stanley has found in the great Equatorial forest … The two tribes of savages, the human baboon and the handsome dwarf, who will not speak lest it impede him in his task, may be accepted as the two varieties who are continually present with us – the vicious, lazy lout, and the toiling slave.

In *How the Poor Live*, George Sims (1976, pp.64–5; first published 1883) drew the same parallel. There exists 'a dark continent that is within easy walking distance of the General Post Office … the wild races who inhabit it will, I trust, gain public sympathy as easily as [other] savage tribes.' In *London Labour and London Poor* (published 1861), Henry Mayhew (1967; see also Humphreys 1977) drew a related comparison between civilised, stationary 'races' and London's 'wandering tribes' of 'Bushmen' and

'Hottentot Sonquas' (for discussion see Himmelfard, 1971). The only occasion when Mayhew named any of his working-class subjects as white is also worthy of mention. For it is marked both by an inference of racial unreliability and by its displaced colonial context. The reference comes in a passage depicting 'Negro beggars' on the streets of London. In the same section, Mayhew mentions 'white beggars'' emulation of these 'negroes'' particular practice of pretending to be ex-slaves. '[M]any white beggars,' Mayhew writes, 'fortunate enough to possess a flattish or turned-up nose, dyed themselves black and "stood pad" as real Africans' (1950, p.391). The account demonstrates more than 'white beggars'' physical proximity to blacks. It also carries the suggestion that their whiteness is perfunctory, that it is merely a logical function of the imaginative transposition of a colonial scenario onto Britain (an imaginative process triggered as a consequence of 'negroes' living in London) and that it carries none of the connotations of superiority and civilisation accorded to hegemonic conceptions of white identity.

It is also noticeable, in mid–late Victorian commentary, how often comparisons are drawn between the social roles of the non-white subjects of colonial and settler societies and the white working class. Malik (1996) has recently examined the way black people and the English working class were routinely understood by reference to each other. Thus he cites Edwin Hood's observation that 'The negro is in Jamaica as the coster-monger is in Whitechapel; he is very likely often nearly a savage with the mind of a child' (p.97). Malik (p. 93) also quotes an overview of working-class life offered in an 1864 edition of *The Saturday Review*: 'The Bethnal Green poor ... are a caste apart, a race of whom we know nothing, whose lives are of quite different complexion from ours.' Drawing on a US model of race relations, the article continues, 'distinctions and separations, like those of the English classes which always endure, which last from the cradle to the grave ... offer a very fair parallel to the separation of the slaves from the whites.' Lorimer (1978) also draws attention to such depictions. He cites *The Daily Telegraph* (from 21 August 1866) as referring to white working-class rioters as 'negroes':

> There are a good many negroes in Southampton, who have the taste of their tribe for any disturbance that appears safe, and who are probably imbued with the conviction that it is the proper thing to hoot and yell at a number of gentlemen going to a dinner party.

The *Daily Telegraph*'s diatribe was not a case of mistaken identity but rather a self-consciously ironic application of an increasingly influential metaphor of social difference, namely colour. The focus on the aggressive nature of the urban working class evidenced in the passage also suggests that the 'darkening' of the British working class may have been encouraged by the development of middle-class fears about the rise of a politically rebellious, foreign-influenced, proletarian culture. Thus we find that, although no section of the working class was actively included in whiteness, the most actively excluded were the urban working class, or 'this new city race' as Charles Masterman termed them in *From the Abyss* (1976, p.242; first published 1902). The 'new race' or 'city type' was also depicted in Masterman's *The Heart of the Empire* (1901). This was 'a generation knocking at our doors', a 'stunted, narrow-chested' rabble, 'yet voluble, excitable, with little ballast' (p.8).

Clearly, the racial othering of the working class was not always simply a matter of class. The middle class sense of alienation from, and fear of, this group was also centred upon their metropolitan nature and the place of immigrant labour within their ranks. Before returning to the distinction between urban and rural, I shall briefly address the latter theme.

Concern with 'immigrant' groups in nineteenth-century British cities drew on an extensive and, in part, well-established repertoire of xenophobic and racist categorisations of 'other Europeans' as, metaphorical or literally, less white than the natives. However, the group that attracted the most opprobrium were migrants within the British Isles, namely the Catholic Irish. The late nineteenth century saw a hardening, an increased literalness, of interpretations of the Catholic Irish as racially other than, and as darker in complexion than, other Britons, more especially the English. The scientific identification of the 'so-called black Celts' (Huxley, 1979, p.167; first published 1870) or 'dark Euskarian type' (Allen, 1979, p.241; first published 1880) amongst the Irish population gave credence to pre-existing mythological genealogies. Thus, for example, in an analysis that draws the representation of the Irish and Jews together with that of Africans, Cohen (1988, p.74; see also Innes, 1994) notes that 'the special name given to the Irish – Milesians – exemplifies the transgressive role assigned them in English constructions of whiteness. Milesians, he explains, were 'descendants of the legendary Spanish King Liesus. This "race" was supposedly the product of interbreeding between Moors and marrains, that is between Africans who had Hispanic roots and Spanish Jews ... This hybrid status did more than confirm the Irish as a "monstrous race". It set them apart, made them a special case.' The trend towards literalising the darkness of the Irish may be exemplified by the anthropological investigations of T.H. Huxley (1894) and John Beddoe (1885). Huxley drew a distinction between 'the dark whites' and 'the fair whites' of Britain. The former he termed 'Melanochroi', a group he considered to comprise 'a broad band stretching from Ireland to Hindostan' (p.262). The latter he called 'Xanthochroi', a people that 'continues in force' from England 'through Central Europe, until it is lost in Central Asia' (p.261). Although Huxley designated the Melanochroi and Xanthochroi as separate races he claimed both to be equally civilised. In this respect his research may be contrasted with the work of Beddoe, the founder of the Ethnological Society. For a Victorian audience the inferences of Beddoe's 'Index of Nigrescence' – a quantitative measure based on hair colour which he applied across the British Isles – were clear: that the darker-skinned races of Britain derived in some way from 'Negroes' and, hence, were inferior. Indeed, noting a 'greater tendency towards melanosity' (Beddoe, quoted by Curtis, 1997, p.20) within western, Celtic areas of Britain and Ireland, Beddoe coined the term 'Africanoid' for certain Welsh groups and peoples from west Ireland. In his analysis of cartoons of Irish rebels in Victorian comic weeklies, Curtis (1997) allies

Beddoe's analysis to a general simianisation of the Irish in Victorian culture. Disputing Foster's (1993) account in *Paddy and Mr Punch*, Curtis makes a point of asserting that the representations of non-white Britons found in such magazines signified racial not class division in Britain. '[P]lebian *Englishmen* with apelike features,' he argues in the revised edition of *Apes and Angels: The Irishman in Victorian Caricature*, 'were very much the exception to the rule in English comic weeklies … English comic artists, in fact, saved their best simianising efforts for Irish rebels' (Curtis, 1997, p.121). However, Curtis's singular focus on simianisation in comic weeklies should warn us against extrapolating his observations onto all Victorian representations of the working class. Comparisons with apes were not the only, or necessarily the most important, way non-whiteness was connoted. The material I have already cited from the Victorian 'urban explorers' is suggestive of the range of similes and metaphors used to distance and define the poor as the dark other of bourgeois whiteness. Indeed, it is far less the race of the working class that is noticeable from this material than their geographical location: it is the *urban* poor that are excluded from whiteness.

As a consequence of the large-scale migration of rural families to the city that characterised the late nineteenth century, the British (and, again, more especially the English) working class was often construed to be losing its national and racial rootedness. 'Traditional' rural folk were being lost to racial degeneracy (Rich 1994). In *Rural England*, Rider Haggard (1976, p.218; first published 1902) noted that this migratory flow 'can mean nothing less than the progressive deterioration of the race' (see also Bray, 1907; Inge, 1919). In *The Poor and the Land*, the same author contrasted the 'puny pygmies growing from towns or town bred parents' with the 'blood and sinew of the race', the 'robust and intelligent' countryman (Haggard, 1905, p.xix). The racialised contrast between urban and rural relied, in part, on an association of the urban with immigrant labour. However, the perceived threat of the urban also drew upon an existing and specifically English discourse of national and racial romanticism which placed the essence of Englishness in 'the people['s] … natural breeding and growing grounds' (Lord Walsingham, quoted by Low, 1996, p.19), the countryside. Indeed, the similarities between the stereotypes of the 'rosy cheeks' and 'healthy complexion' of the English peasant and the 'vigorous' nature of bourgeois whiteness are suggestive of Victorian middle-class writers' investment in rural nostalgia as a kind of origin myth of their own ascendance. This impression is strengthened by the fact that rural workers in the late nineteenth and early twentieth centuries tended not to be subjected to horrified explorers but rather to reverential cultural retrieval. By 1911, folklore studies had been published on twenty-nine of England's forty counties. Colls's (1986; see also Howkins, 1986) observations of the development of 'folk study' in the period draw a direct contrast between the valued purity of the rural past and the racially degraded urban present. Citing the 'Warwickshire peasant' Joseph Arch, a figure introduced 'to the nation' by the Countess of Warwick, Colls notes that

When the black American 'cake walk' dance was introduced to the London music halls [in 1898], so dreaded was miscegenation with 'old, healthy sensual (but not sensuous) English dances' that cake walking was said to show 'why the negro and the white can never lie down together'. That the South Londoners had mixed cake walking with their own swagger to dance the first Lambeth Walk in 1903 was so much the worse for them. (Colls, 1986, p.47)

In describing the urban working class, the use of the language and logic of race became so commonplace that it may be argued to have formed a new and distinctive kind of racialised discourse. In other words, the original point of reference for the association of the working class with 'blacks' or 'browns' seems to have faded as the former's racialised othering took on a linguistic and socioeconomic life of its own. Not unrelatedly, a transition, or slippage, from race used as a metaphor of otherness to race used as an objective description of working-class difference can be witnessed. The working class were increasingly not simply *like people* who were different – *they were* different. By the late nineteenth century, it had become increasingly common for social and scientific theorists to use race as a kind of general principle of hierarchy. Thus, for example, Count Gobineau (1966, p.120) noted (in the influential English translation of *The Inequality of Human Races*, published in 1915), that

> every social order is founded upon three original classes, each of which represents a racial variety: the nobility, a more or less accurate reflection of the conquering race; the bourgeoisie composed of mixed stock coming close to the chief race; and the common people who live in servitude or at least in a very depressed position.

In similar vein, Gustav LeBon (1912, p.29), another translated French writer, noted that 'The lowest strata of the European societies is homologous with the primitive men.' In time, LeBon (p.43) continued, 'the superior grades of a population [will be] separated from the inferior grades by a distance as great as that which separates the white man from the negro or even the negro from the monkey.' In Britain, the founder of eugenics Francis Galton (cited by Malik, 1996, p.99; see also Darwin, 1901; cf. Galton, 1979; Pick, 1988) opined that he found it hard to distinguish 'the nature of the lower classes of civilised man from that of the barbarian'. Galton continued that 'classes E and F of the negro may roughly be considered as equivalent of our C and D'.

I have sought to show that working-class Britons in the Victorian era (most especially the urban and non-English working class) were marginal to the symbolic production of white identity. My sources have been derived from the bourgeois class, from men who saw a profound gulf between themselves and the people of the 'dark' regions upon which they speculated. In the next section, my sources will also derive from those who situate themselves, and their audience, at the centre of white identity. Yet, as we shall see, these are different voices, not bourgeois voyagers but exponents of 'ordinary people'.

Symbolising the ordinary: the transformation of whiteness in Britain

From being marginal to whiteness in the nineteenth century, the working class came to adopt and adapt this identity in the twentieth century. With the rise of mass non-white settlement in Britain it may be supposed that the colonial imagination was rearticulated as a discourse about a 'newly multi-racial' British society. In other words, the working class 'took up' whiteness from the available repertoire of colonial and hierarchical terms in 'response to' the presence of non-white Britons. Although this analysis usefully highlights the colonial heritage of certain ideas about whiteness (especially its continuing association with superiority), it fails to engage with either the particularity of white working-class identity or the history of white identity's location within British class society.

The reformation of the symbolic economies of race and class cannot be abstracted from the reformation of British capitalism. The two processes which I offer as explanations of the transformation of white identity are the formation of populist imperialism and the transition from liberal to welfare, or advanced, capitalism. I shall be suggesting that these two phenomena are related and, more specifically, that the former presaged the latter.

The extent of the influence of ideologies of imperialism on working-class life is the subject of some controversy (Pelling, 1968; Colls and Dodd, 1986; MacKenzie, 1985). It seems certain that their grip on upper- and middle-class culture was more profound, more intimately constitutive of everyday class experience, than for other groups. However, the number of pointers to the cross-class significance of the imperialist and nationalist propaganda disseminated by large swathes of the voluntary and education sector and media/entertainment industries is also impressive. Intriguingly, given the representations of urban darkness previously discussed, for some critics of the period it was precisely the degraded nature of city living that made the working class pliable to the baser forms of nationalism and imperialism. Thus, for example, in his *The Psychology of Jingoism* the liberal critic J. A. Hobson (1901, p.7) asserted that xenophobic nationalism was associated with the 'bad conditions of town life ... lowering the vitality of the inhabitants.' The notion that, when articulated within working-class culture, national and racial ideologies are marked by their irrationality and crude parroting of elite ideologies, finds echoes across a range of early twentieth-century social commentary. This theme of debasement has been challenged recently by Crowhurst (1997; see also Featherstone, 1998), who argues that working-class audiences in Victorian and Edwardian music halls identified with, and represented themselves through, stereotypes of blackness. Crowhurst's contention is interesting but speculative. Far more overwhelming is the evidence that suggests that, from the late nineteenth century, populist nationalist and imperialist activities brought imperialism and racial categories into ever closer proximity with working-class lives. Autobiographies from the period are one measure of this process: for example,

Fred Willis's (1948) account of learning about racial divisions in his Victorian school and Alfred Williams's (1981) memories of the patriotic fervour that dominated the railway factory where he worked. At the end of his study into how military youth organisations, such as the National Service League and the Territorial Force, recruited successfully in working-class Birmingham, Michel Blanch (1976, p.119) concludes that through their work 'imperialist and nationalist sentiment obtained real roots in working class opinion.' The masculinist ideologies and recruiting practices of these organisations may suggest that imperialist sentiment and categories were more actively developed amongst working-class men than women. This gender difference is less apparent, however, in another influence on working-class views on race, the popular press. In his discussion on the historical development of working-class racism Miles (1982), drawing on Street (1975), argues that because of limited literacy and education 'the articulation of racism was probably the prerogative of the dominant class up to and including the nineteenth century' (p.118). However, Miles continues, the introduction of mass-circulation fiction with imperialist themes in the late nineteenth century enabled 'the reproduction of those racist notions of the bourgeois world-view, some of which were, in turn, derived or developed from scientific racism, to an audience of the working class' (p.119).

Nevertheless, the relationship between the changing racial identity of the British working class and imperialism was more than simply a question of the dissemination of imperial and/or bourgeois cultural ideologies. It also reflected imperialism's impact upon the economic and political restructuring of British capitalism. One particular aspect of this latter process that interests me here is the way imperialism introduced significant non-market, state-interventionist and social consensus (most importantly, popularist nationalism) tendencies into British society. This relationship is partly explained by reference to the widespread conviction that welfare was needed in order to breed the kind of healthy and vigorous race capable of controlling an empire. As the leader of the Liberal Party, Lord Rosebery (cited by Semmell, 1960, p.63), noted in 1902, 'The imperialism that grasping after territory, ignores the condition of the Imperial Race is a blind, a futile, and a doomed imperialism.' With the physical and military weakness of Boer War recruits in mind, Rosebery went on to advocate housing and educational reforms capable of nurturing a strong and willing 'imperial race' (see also Semmell, 1960; Cohen, 1985; Stedman Jones, 1971). Such interventionist tendencies inevitably unsettled the established symbolic economies of Victorian free-market liberalism. Government manipulation of socioeconomic conditions in Britain and around the world, combined with the development of a range of media institutions and philanthropic and civic initiatives in British cities, enabled the formation of a truly popular 'national community' (a community in which the standard of living of working-class people and their ideological integration into the nation was enhanced beyond recognition (the economics of this process are detailed by Hobsbawm, 1969, the nationalist and other ideological connotations by Pieterse, 1990)). Imperialism signalled a shift within capitalism, a move away from laissez-faireism and the exclusion of the 'British', 'white' working class from key national–racial symbols and towards an interventionist, social consensus-oriented form of capitalism. In such a society, symbols of social status and images of the national ideal could begin to be adopted and adapted for popular usage on a much greater scale.

The explication of these tendencies in imperialism provides a clear parallel with the more general shift within twentieth-century capitalism towards state interventionism and welfarism. Although this trend has complex roots, two main influences may be abstracted: first, working-class struggle (which may be exemplified by reference to rising trades union membership and industrial militancy; though see Pelling, 1968): and, second, the increasing complexity and consumer orientation of capitalist production and management. Although the former influence appears non- or even anti-capitalist and the latter part and parcel of capitalist reproduction, the interests of both were partially responded to by the expansion of 'national' and 'free' public housing, health, education and 'social security' schemes and benefits. Introduced fitfully throughout the century, and developed most fully in the immediate post-Second World War period, state welfare helped to produce a population ideologically committed to, and capable of participating in, 'state-managed capitalism'. Thus welfarism both fused and recuperated contradictory and potentially explosively antagonistic social forces into a national project (see London Edinburgh Weekend Return Group, 1979; CCCS, 1981; see also Offe, 1984; 1985). As this implies, the benefits of welfare were articulated as a national, and nationalistic, discourse. Welfare came wrapped in the Union Jack.[3]

As the chasm of class identities apparent in the Victorian period was narrowed, the marginalisation of the working class from whiteness became untenable. In this new British social formation, racial and national identities once centred on the elite become available to the masses. These identities were also able to be *adapted* by the working class. As this implies, the symbolic reconstitution of capitalism saw changes within both the social boundaries of whiteness and its connotative range. As we saw earlier, in the late eighteenth and nineteenth centuries whiteness connoted an idealised human being, a figure depicted in terms of its *extraordinary* qualities. These connotations, symptomatic of the bourgeois class meaning of whiteness, continue to influence twentieth-century racial discourse, including the racial discourse of the white working class (see, for example, Rose *et al.*, 1969). However, they have entered into a – sometimes difficult – relationship with another set of associations, associations that reflect the new class politics of white identity. For now whiteness, as well as being a supremacist identity, is cast as the identity of the ordinary; it evokes a lack of exceptionality, the homely virtues of quietness, tidiness, cleanness and decency. Thus the connotative emphasis of whiteness has been mutated and broadened, from being a virtually superhuman identity to being, at least in part, a prosaic one. Both sets of associations privilege whiteness and reflect its racist implications. However, they do so in different ways and, at least within Britain, are indicative of different forms of capitalist class relations.

To exemplify this connotative reformation, I will turn to a variety of assertions of white working-class identity articulated by white, mostly working-class, men from the 1950s through to the 1970s. The examples offered all share an aggressively defensive position against perceived challenges to white people's social and economic status. I would stress that they are not designed to be a representative sample of 'white opinion'. What they do provide are clear expressions of the political mobilisation of whiteness, of whiteness being 'put to work' in order to cohere a racialised community of solidarity and resistance. I shall, first, note the relation between whiteness and welfarism and nationalism and, second, the values ascribed to white identity.

Mid–late twentieth-century British racist discourse is often character-ised by its allusions to 'the national community' (within England, an imperialistic conflation of 'England' with 'Britain' is also often apparent). Moreover, such constructions frequently posit colour as the key site of exclusion and inclusion. This narrative has required the development of a new history of Britain/England, a narrative of a 'white country' defend-ing itself against – as one British National Party candidate put it in 1964 – 'coloured immigration into our ancient land' (Bean, cited by Woolcott, 1965, p.40). A similar narrative is cited by Phizacklea and Miles.

Q: Why did you vote National Front?
A: Because our dads fought to keep this country a white man's country ... When I first went in [i.e. joined the National Front] I had visions of a united England, you know, England for the English. I know it's racist, but it appealed to me. (cited by Phizacklea and Miles, 1979, p.115)

It is interesting to observe how often such appeals to a cross-class racial–national community evoke welfare structures. The use of the public edu-cation, social benefit, health and housing systems by non-white 'immi-grants' appears to be deeply resented by many white Britons (see, for example, Rose *et al.*, 1969; Runnymede Trust, 1991). Sidney Jacobs's (1985, p.14) analysis of this attitude in the area of housing highlights the 'repeated variations of the accusation that "they've come to sponge off us, taking our jobs, our houses", [an attitude that] contained the demand that black people be denied rightful access to state welfare provision.' Con-firming Jacobs's thesis, Rose *et al.*'s (1969, p.580) survey of white attitudes to the letting of council houses found that

[o]verwhelmingly the attitude expressed was that access to council dwellings should be limited to "our own people" that it was part of the benefits of the Welfare State that should not be shared.

The testimonies gathered by Rose *et al.* make it clear that 'our own peo-ple' means white people. Welfare structures are posited as 'ours'; they are implicitly or explicitly presented as symbolic and material rationales for working-class identification with the national–racial unit. Non-white immigration is thus portrayed as a threat to working-class 'gains'. 'Immi-gration has dragged us back twenty years,' opined the Labour deputy mayor of Deptford, a working-class London borough, in 1964 (cited by

Sherman, 1965, p.111). '[I]t's all right talking about the brotherhood of man, but our first job is to defend the gains we fought for here.' Such allusions to 'our' welfare state also have a central place in the ex-Conservative minister Enoch Powell's oratory on race and nation.

> while to the immigrant entry to this country was admission to privileges and opportunities eagerly sought, the impact upon the existing population was very different ... they found themselves made strangers in their own country, they found their wives unable to obtain hospital beds in childbirth, their children unable to obtain school places. (1968, p.185)

Once one accepts that mid- and late twentieth-century articulations of white identity posit, and are enabled by, a consensual and welfare-oriented capitalist social formation, one begins to hear echoes of this process reverberating throughout British popular culture. At the same time, certain contradictory political dynamics come into view. Most significantly, that welfarism represents a partial achievement of class struggle but also enables and is sustained by a racialised vision of the national community. Unemployment benefit is another arena where this contradiction is played out, the labour exchange or social security office becoming a site of racialised class politics where the 'rights' of Britons, the working class and white people are conflated. The latter point is expressed in the remarks of one white youth, identified as a member of the youth cult known as 'teddy boys', interviewed by a BBC reporter in the immediate aftermath of the 1958 Notting Hill 'race riot'. Asked to explain the reasoning behind similar white youths' attacks on West Indian immigrants, he simply says the following:

> I mean down the labour exchange, I mean you go in there, we were in there today, three of us, and we're standing waiting for our pay and we see one darkie go up, he drew eight – seven pounds something, we about six whites go up, draw three pounds, another darkie go up and draw eight pounds and so forth and so on. And it's nearly all blacks down there except for the few whites. Spades, I'm sorry – spades. (in Glass, 1961, p.264)

This particular respondent was interviewed amongst a group of similarly racist white working-class youths. Their attitudes to race also displayed the other characteristic of mid–late twentieth-century white identity that I mentioned above, its symbolic association with the 'ordinary'. Whilst white identities have been used normatively since their inception, their adoption and adaptation by the British working class has seen an assertion, less of whiteness as a universal ideal, the peak of humanity, and more of whiteness as 'ordinary' and 'decent'. The teddy boys' complaints about non-whites are uniformly prosaic, hinging on accusations of being 'filthy', not having 'good manners' and of 'causing scandal'. Indeed, the perpetration of racist violence, which the teddy boys readily admit to, is justified by another interviewee through reference to the most seemingly banal transgressions of 'polite behaviour':

> You go on a bus – to a white conductor you'll ask for a tuppenny fare, they'll say thank you – you go on one with a darkie, you ask for a fivepenny fare, he'll say fares please and just walk away. They're as ignorant as I've ever met, and I'll have a go with them any day they fancy. (in Glass, 1961, p.264)

Similarly, Enoch Powell's 'Rivers of Blood' speech, made in 1968, which called for the repatriation of 'immigrants' (i.e., non-whites), repeatedly alludes to the concerns about immigration of the 'decent, ordinary fellow' (p.180), the 'ordinary, decent, sensible people' (p.186). Woven into this rhetoric are allusions to another and, within the English context, more immediately recognisable form of class symbolism, namely, respectableness. The 'respectable street in Wolverhampton' beset by 'immigrants' that Powell uses to exemplify his general message is designed to evoke images of a white working class assailed by non-whites. Indeed, the shift from white to non-white occupation of the street is symbolised as a move from the ordinary to the exceptional, the humdrum to the extraordinary, 'The quiet street became a place of noise and confusion. Regretfully ... white tenants moved out' (p.187).

So far in this chapter, I have sought to show how white identity, from being incorporated into the reproduction of Victorian laissez-faire capitalism, became integrated into consensus-oriented, state-managed capitalism. The racist commentaries cited in this section exemplify a shift in the symbolic constitution of this identity, a shift linked to the transformation of British capitalism. However, welfarism is not necessarily the end point of capitalist development. The past couple of decades have witnessed a concerted 'rolling back' of the state in certain areas and the 'reduction of "welfare state" to "social insurance"' (Roebroek, 1992, p.66) in others. Bauman (1997) has highlighted signs of a transition in popular constructions of welfare. From something that is 'ours', a common good, welfare structures have, he argues, increasingly been cast as something that 'we' give to 'others'; in other words, an altruistic burden. He speculates that,

> On the way from Beveridge-style social insurance as universal entitlement to means-tested handouts for 'those who need them most', there must have been a critical point, difficult to spot in time, yet quite real, at which the 'community' turned in the eyes of the majority from a safeguard of individual security into a noxious drain on individual resources; from something 'that is our right' to something 'we cannot afford'. (p.5)

Extending this train of thought, it is tempting to speculate on a shift in the racial meaning of the plural pronouns that Bauman places in inverted commas. From welfare being something the white national–racial unit awards to itself, it has become something that that same unit awards to, and associates with, 'others' (i.e., ethnic and other marginalised minority groups). The temptation within this analysis is to imagine a return to a Victorian class–race order, with both visible minority groups and the white underclass being cast as a 'new rabble' (Murray, cited by Piachaud, 1997, p.3), a congenitally inferior and literally and metaphorically 'dark' section of society. Writing this, I am reminded of the phrase I have heard used on several occasions to describe two of the virtually all-white, and highly impoverished and welfare-dependent, areas of my own city, Newcastle, namely 'darkest Benwell' and 'darkest Scotwood' (see Nayak, 1999, for discussion; see also Campbell's 1995 account of Benwell as 'little Beirut'; also Osmond, 1995; Nayak, 1999).

However, while these latter phenomena require our attention, they remain, as yet, only tendencies. Put simply, it appears, at present, that welfare capitalism is being reorganised but not scrapped. '[M]ore (or rather less) of the same' is how Taylor-Gooby (1989, p.639) summed up those changes apparent by the end of the 1980s. Ruggles and O'Higgens (1987, p.187) concurred, observing that even New Right-dominated governments 'have moved to acceptance of the structural role of welfare in democratic mixed economies and are now focusing on ways of constraining and redistributing its cost.' Welfare capitalism is still with us, if for no other reason than, as Offe (1984, p.153) put it, 'The embarrassing secret of the welfare state is that while ... capitalism cannot exist with, neither can it exist without, the welfare state.'

Thus I would conclude that, although the scale of contemporary reorganisation, and the emergence of socioeconomic gulfs comparable with those of Victorian Britain, may suggest the potential for movement within the symbolic constitution of whiteness, it is premature to announce the death of racialised welfare capitalism. Not unrelatedly, whiteness remains, at present, available as a mass identity. It is something claimed by, and claimed for, all European-heritage Britons. It may, one day, retreat back to an elite identity or, more hopefully, lose its connotative power and fade into insignificance. One day maybe, but not yet.

'White studies' or 'American studies'? A concluding remark

The last ten years have seen the development of a historical and sociological literature on white identities. The vast majority of this work has emerged from, and is focused upon, the United States of America. Noel Ignatiev (1997), a leading exponent of this new school, recently noted that his work was only applicable to the latter country, and that its explanatory framework should not be imposed on other societies. Unfortunately, the analytical categories and concerns generated within North American debate have an international influence beyond the control of their progenitors. Indeed, the imprint of American models of racial identity may today be seen across the globe, from indigenous rights campaigners in Japan aligning themslves with Black Power and the American Indian Movement (Siddle, 1997), to the establishment of 'branches' of the Ku Klux Klan in southern England. This globalising momentum demands, I would suggest, the homogenisation and 'Americanisation' of the world's diverse racial histories. Thus, for example, representations of the struggles of racialised minorities in a wide variety of countries increasingly resemble the archetypes and stereotypes generated within and around African American and Native American history. I am not concerned here to evoke indignation at this process but rather to suggest that it is worthy of consideration and study and that it obliges scholars to apply a degree of circumspection to the limits and implications of current American histories of whiteness. Indeed, the value of this latter body of work will be considerably diminished if its geographical particularity, its localness, is not recognised. The study of the development of white identity amongst working-class Britons necessarily includes elements (especially the role of imperialism and welfarism), as well as a time-scale, not be found within the work of Ignatiev, Roediger and Allen. This is not to make a claim for a unique British 'national experience' – many of the themes narrated here may also be found within other European traditions of whiteness – but rather to explain why current discussion on whiteness needs to be internationalised, to be refocused as a global debate and not simply an American one.

1. In *Colour, Class and the Victorians*, Lorimer (1978, p.206) notes 'The popular stereotype of the Negro in the mid-nineteenth century owed more to the New World than to Africa. Even then American rather than West Indian images predominated in anti-slavery rhetoric, in the popular fiction of Harriet Beecher Stowe, and in the minstrel shows. The mid-Victorians had become so familiar with the Negro in these contexts that when an imperial crisis did impinge upon their consciousness, as from Jamaica in 1865, the event made few new impressions and simply confirmed existing viewpoints.'

2. Stowe's contention provoked widespread dispute in Britain. For example, *Uncle Tom in England: or a Proof that Black's White* (1852) sought to show that American slavery and the British class system could not be regarded as in any way equivalent.

3. It is pertinent to recall that *The Beveridge Report* – often seen as the founding text of the post-Second World War welfare state – offers, as Cohen (1985, p.88) notes, 'an explicit incorporation of pre-war assumptions of efficiency and eugenics'. Amongst the paragraphs in the report Cohen highlights is the assertion that 'with its present rate of reproduction the British race cannot continue', hence, 'means of reversing the recent course of the birth rate must be found.' Beveridge's conception of the welfare system was also encapsulated in his essay 'Children's Allowances and the Race' (cited by Cohen, 1985, p.89):

> Pride of race is a reality for the British as for other peoples … as in Britain today we look back with pride and gratitude to our ancestors, look back as a nation or as individuals two hundred years and more to the generations illuminated by Marlborough or Cromwell or Drake, are we not bound also to look forward, to plan society now so that there may be no lack of men or women of the quality of those earlier days, of the best of our breed, two hundred and three hundred years hence?

Aneurin Bevan, another founding father of the welfare state, is also cited by Cohen (p.89) proclaiming, as minister of health in 1949, that he had 'arranged for immigration officers to turn back aliens who were coming to this country to secure benefits off the health service.'

A white world? Whiteness and the meaning of modernity in Latin America and Japan

Introduction

> [W]e have always been looking at Western countries as progressive
> ones. These were places that Japan had to catch up with. From this there
> developed sort of a complex – 'it's a white world'. (Creative Director,
> Dentsu advertising agency, Tokyo, cited by Creighton, 1997, p.216)

The notion that 'it's a white world' appears both obvious and faintly
ridiculous. The idea that those social, economic and cultural forms that
dominate the planet may be characterised as white, whether in terms of
their origin, their values or which group benefits from them, is a ubiquitous one. Yet it is frequently swiftly followed by certain caveats – 'for the
time being at least'; 'or so they like to think' – qualifiers that simultaneously undermine and highlight the pomposity of the initial claim. The
demands and delusions of whiteness are keenly observed by many people around the world, its conceits identified, scrutinised and often
resented.

One of the most important ways the idea of white superiority has been
articulated and reproduced is through the assertion that whiteness is
coveted by non-whites because it allows them entry into modernity, in other
words, that people without European ancestry beg, borrow or steal the
ways of whiteness in order to achieve, or rather *dissemble*, 'developed',
'advanced' social, political and aesthetic standards. Henry Champley
(1936), the French journalist I introduced in Chapter 1, offered a number
of observations on this process. Throughout *White Women, Coloured Men*,
Champley exhibits the mixture of flattered bemusement and contempt
that often accompanies white interest in the attempts of others to
become more like them. In particular, non-white women's desire to ape
white women, to become what Champley calls 'fake-white' women, is
recorded with smiling condescension (Figure 3.1). Thus, for example,
Champley tells us that, whilst he was resident in China, young Chinese
women 'asked me to teach them our modern dances' (p.175). He
explains that the skills of the waltz and the tango that he passed on were

Figure 3.1 'A Chinese Cinema Poster showing a "fake-White" woman' (original caption), from H. Champley, *White Women, Coloured Men* (1936)

employed by the natives in order to deceive, in order to establish an inauthentic identity, a mischievous claim to being modern. Reporting the popularity of 'white dances' and 'white manners' amongst 'Coloured women' throughout East Asia, Champley suggests,

All these Coloured women clearly had but one ideal, which was all the rage: to look like White women. Some of them achieved nothing but a sickening caricature. But others, above all the half-breed girls, sometimes created the disturbing effect of a regular double. Two or three were extraordinary – more White, more civilised, than real White women. (p.176)

Champley was writing in the 1930s. But, shorn of his overt racism, one can easily imagine similar remarks being made at the turn of the twentieth century. The notion that non-whites, particularly non-whites within non-Western societies, seek to imitate white attitudes, clothes, entertainment, design and so on remains a popular topic within Western journalism. From excited accounts of 'eye-widening' operations in East Asia to inflated commentaries on the global influence of the English language and Western pop music, there appears to be a ready market for stories on Western culture's manifest destiny to sweep all before it, more specifically on how 'they' want to become more like 'us'.

The aim of this chapter is to assess, hopefully with a little more care than exhibited by Champley and his ilk, the adoption and adaptation of white identity in societies commonly regarded in the West as non-white. I shall be arguing that racial whiteness does provide a powerful influence on cultural and economic history around the world. However, this phenomenon cannot be explained by reference to some quality innate in European racial whiteness. Indeed, it only becomes comprehensible when the power of whiteness as a key symbol within modernity is identified. Moreover, far from simply imitating whiteness, it will be shown that whiteness is creatively mutated, that it can become deployed in original ways and for diverse political reasons.

The notion that Western economic and social influence has expanded across the Earth is a leitmotif of nearly all studies of post-fifteenth-century global change. Indeed, the intrusion, as well as the assimilation, of Western-identified ideologies and practices might be said to form the core problematic of modern history. Although many aspects of this process have been extensively debated, the way these ideologies and practices were and are racialised remains relatively undiscussed. For European social and economic paradigms were connoted through the symbols of race, symbols that gave capitalist incursion and modernity a European, and hence white, identity. It follows that the history of this era may benefit from the consideration of the relationship between whiteness and modernity. More specifically, that any understanding of the interpretation, or translation, of 'Western modernity' into different cultures around the world demands an understanding of these cultures' adoption and adaptation of notions of racial whiteness.

Thus this chapter introduces what may be termed the 'whiteness of modernity' in the non-white world. The scale of this topic demands that I am selective with my sources and clear as to the limits of my conclusions. I shall be highlighting just a few instances of the reception of, and engagement with, whiteness: instances that are framed by particular national narratives. More specifically, my sources will be drawn from what tend to be regarded in contemporary Europe and North America as non-white societies, namely Latin America (more specifically Brazil and Venezuela) and Japan. These examples should not be read as generalisable to other non-Western countries. Although broader themes on the place of whiteness in contemporary capitalism do emerge in this discussion, to be either accurate or instructive, the histories that will be discussed here need to be accorded their geographical particularity.

My concentration on the socioeconomics of the interpretation of whiteness may be contrasted with a more familiar way of writing about the impact of whiteness on non-white people, namely analysing the *psychological damage* caused by white racism within the non-white psyche (a tradition that includes Du Bois's *The Souls of Black Folk* and Fanon's *Black Skin, White Masks*). This latter approach has exhibited a strong tendency to evoke the theme of 'racial misrepresentation': to produce assessments of the problem of white distortion and destruction of the personal and social identity of other races. Although these kinds of psychological impact are clearly important they do not form the focus of the present chapter. My emphasis is upon how whiteness works as a process of social change, a process that reflects and shapes other dominant socioeconomic discourses.

The following account is divided into two. In the first section, I address Latin America, focusing on the history of whitening in Brazil and Venezuela before turning to a very particular contemporary phenomenon, namely the appeal of the white, blonde Brazilian media 'megastar' Xuxa. In the second section, I introduce material from East Asia, more specifically I address the history of the adoption but also the assimilation – what might be called the strategic reading and appropriation – of whiteness in Japan.

The white modern in Latin America

> Anybody who watches Brazilian television for half a day sees that it is dominated by whites and by white images of power, success, intelligence and beauty. (Simpson, 1993, p.38)

The power and prestige of the European colonising powers in Latin America was interpreted through and justified by categories of identity. Amongst the most important of these identities were the associated terms 'Christian', 'civilised' and 'white'. These themes were deployed from the late sixteenth century to depict and legitimise Portuguese and Spanish conquest throughout the Americas. More specifically, they were employed to organise the economic and social structure of colonial societies. The most inflexible of these identities, the one that was least available to be assimilated either by native people or by imported African slaves, was whiteness. Not unrelatedly, the possession of whiteness became the key symbol of access to economic and social status. Writing in the early nineteenth century, the German explorer Alexander von Humbolt drew attention to the specifically colonial context of claims to white identity:

> In Spain it is a kind of nobility not to descend from Jews or Moors. In America, the skin, more or less white, is what dictates the class that an individual occupies in society. A white, even if he rides barefoot on horseback, considers himself a member of the nobility of the country. (cited by Mörner, 1967, pp.55–6)

The further away from whiteness one was understood to be the more difficult it was to obtain either wealth or respect. Indeed, a complex system of racial classification arose in most countries in the Americas that illustrated both 'the almost pathological interest in genealogy that is characteristic of the age' (Mörner, 1967, p.59) and the role of whiteness as the core of racial hierarchy. Thus, for example, Steven's account, from 1825, lists twenty-three 'mixture[s] of the different castes, under their common or distinguishing names', and their associated colour (cited by Pratt, 1992, p.152). Steven's list includes 'Mestiso' ('White' father and 'Indian' mother; colour described as '6/8 White, 2/8 Indian – Fair'); 'Mulatto' ('White' father and 'Negro' mother, '7/8 White, 1/8 Negro – often Fair'); 'Zambo' ('Negro' father and 'White' mother, '4/8 White, 4/8 Negro – dark copper') and 'Quarteron' ('White' father and 'Mulatto' mother, '6/8 White, 2/8 Negro – Fair').

By the late nineteenth century, the category 'white' was becoming firmly enmeshed in the capitalist transformation of Latin America. It was also increasingly being viewed through the optic of European Social Darwinism. Two interconnected assumptions arose from these new ideologies: the position of 'white' as the top slot in any objective classification of the natural units of society and the association of white people with the virility and dynamism of bourgeois economic development. As we shall see in the ideologies of national 'whitening' discussed below, some of the clearest signs of the role played by white identity in the process of capitalist and national modernisation may be found in the late nineteenth century and first half of the twentieth. However, the independence movements of the early and mid-nineteenth century also evidence the deployment of 'Europeanness' as part and parcel of a recognisably modern(ising) project. The cultural and racial ambitions of the creole (i.e., European-heritage) elites who dominated the independence movements have been summarised by Mary Louise Pratt (1992, p.175) in the following terms: 'the liberal Creole project involved founding an independent, decolonised American society and culture, while retaining European values and white supremacy.' The potential tensions evident within such a 'project' were cohered and concealed within the concept of 'the modern nation'. In other words, they were conflated within a discourse of national independence that offered the prospect of creating new and dynamic European civilisations unshackled by ties to the outmoded and oppressive institutions of old Europe.

As in Europe itself, fantasies of capitalist modernity in Latin America were engaged and disrupted by romantic and pastoralist representations of national identity. Indeed, a contrast between capitalist, industrial Europe and non-capitalist, rural visions of Latin America was deployed as a way of defining the distinctiveness of American society by early advocates of independence (for example, the Venezuelan poet and states-man Andrés Bello). As this implies, to be non- or anti-European, *and hence non- or anti-modern*, has also sometimes played a part in the construction of national identities in Latin America. However, Pratt (1992) draws our attention to the way such symbolic opposition to Europe was always *also* a form of Europeanisation in the service of creole power. She suggests it represented the new elite's adoption of distinctly European categories and conceits of resistance to capitalist alienation. The naturalist perspec-tive associated with von Humbolt was, Pratt notes, particularly signifi-cant in providing creole elites with a way of depicting Latin America as an empty canvas, a vast and untrodden landscape awaiting their control and civilising influence:

> Humbolt's 'aesthetic mode of treating subjects of natural history' re-enacted an América in a primal state from which it would now rise into the glory of Eurocivilization. In the myth that followed from his writ-ings (and for which Humbolt must not be held solely responsible) América was imagined as unoccupied and unclaimed terrain; colonial relations were offstage; the European traveller's own presence remained unquestioned. (Pratt, 1992, p.181)

Pratt's reading of colonial relations is suggestive of two things about the development of white identities in Latin America. First, that we cannot think about this process as simply reflecting the imposition of 'Western values' on 'non-Western' societies: whiteness was actively interpreted and translated. Second, that the limits to non-Western adaptations were set by the power relations within the newly independent countries of Latin America as well as between Latin America and Europe (and North America): 'Western values' may have been 'actively interpreted', but it was the West that had the power to supply the privileged discourses of identity and Western-identified people (namely, the creole elite) who did most of the work of 'translation'.

Money whitens. If any single phrase encapsulates the association of whiteness and the modern in Latin America this is it. It is a cliché formu-lated and reformulated throughout the region, a truism dependent upon the social experience that wealth is associated with whiteness, and that in obtaining the former one may become aligned to the latter (and visa versa). The social or economic nature of racial categorisation in much of Spanish- and Portuguese-speaking America has often been contrasted with the more exclusionary, biologically determinist, notion of whiteness supposedly characteristic of the USA. However, differences between the methodologies of racialisation in the north and south of the continent should not blind us to the fact that whiteness has been sustained from the Arctic Circle to Tierra del Fuego as a key to, and symbol of, social and economic ascendancy. The boundaries of racial status may be more per-meable in Latin America but, as numerous historical and contemporary studies of racial attitudes and racial politics south of the Rio Grande have shown (see, for example, Skidmore, 1974; Wright, 1990; Twine, 1998), whiteness remains the most important element in the organisation of racial identity.

The 'white ideal' is apparent at various levels of society in Latin America. Its power is felt within macro and micro socioeconomic organisation as well as within the sphere of everyday cultural interaction. A typical contemporary case study of this power within the social and economic realm may be found in Hugo Nutini's field work in Mexico. 'Upward mobility has been fluid' in Mexico, Nutini notes,

> but always underlined by racial whiteness as an advantage in the social categorisation of phenotypes: that is the overwhelming majority of the lower classes are phenotypically Indian, while the great majority of the upper classes are phenotypically European. Not even the 1910 Revolution was able to address this fundamental fact of Mexican society, as the social, ruling, and political classes of the country are still primarily constituted by European and light Mestizo phenotypes ... even though Mexicans are overwhelmingly Indian, it is essentially a European country culturally. For example, ideals of physical appearance and standards of beauty are essentially European ... This is easily confirmed by the predominance of European phenotypes that dominate television and all forms of graphic advertising. (Nutini, 1997, p.231)

A reflection of the way white 'ideals of physical appearance and standards of beauty' have come to be integrated into the etiquette of everyday life is offered by Roger Lancaster (1991) in his study of attitudes towards skin colour in Nicaragua. Noting the use of the 'polite terms' *'chele'* for white, European-heritage people (*'chele'* is a Mayan word meaning 'blue', denoting the eyes of Europeans) and *'moreno'* for 'black or very brown skin', Lancaster reports the following exchange with one of his respondents. Whiteness in Nicaragua he notes first, is a desired quality, and polite discourse inflates its descriptions of people ...

> Well then, I wondered, how should I respond to Virgil's mother, when she greets me as Chele? Should I say, *'Buenas tardas, morenita'*? 'No, not at all', I was informed. A proper and polite greeting would be: *'Chele, por cariño'* (or, roughly, 'you, too, are white, in my affection'). When I tried this formula out the following day, the results were exactly as [my respondents] had prescribed. Doña Jazmina appeared flattered, and remarked on my mastery of polite conversation. (p.344)

France Winddance Twine's (1998) interviews with inhabitants of the Brazilian town of Vasalia provide further illustrations of the way whiteness affects interpersonal conduct. The problems and opportunities of marrying someone with darker or lighter skin than oneself are raised time again by her respondents. For example, Tônica, an Afro-Brazilian women who has just married a Euro-Brazilian man, explained to Twine that

> I was really thinking of the children. I used to think, 'OK, let me marry a person lighter than myself because if I marry a dark person like myself, [the children] are all going to be dark – the little children.' But I was still thinking of the children right? ... '... if I found a lighter man ... then I will marry a lighter man because then my children will come out prettier.' (p.93)

Here again we see that the permeability of racial categories in Latin America and the ubiquity of interracial relationships should not be confused with the subversion of white supremacism. A similar point needs to made about the praxis and ideology of mestizaje (i.e., the praxis and ideology of racial mixture). An affirmation of the necessity and desirability of racial mixing may be found in the social practice, as well as within the official narratives of national identity, of many South and Central American societies. For some, this orientation may be taken to provide evidence that (1) Latin American is a more tolerant and anti-racist place than North America, and (2) white identity in Latin America lacks a strong normative function. The former interpretation is simplistic but, overall, plausible. However, the latter contention is wrong. Again, it needs to be pointed out that a white norm dominates the relatively fluid system of racialisation in Latin America just as powerfully as it does in the rest of the continent. This point will be exemplified in the discussion of immigration policy and popular cultural practice provided later. However, a useful introduction to the issue may be made through one of the principal elaborators of mestizaje as an official, national ideology, José Vasconcelos.

Vasconcelos was the Mexican minister for education between 1921 and 1924. He is best remembered today as the man who commissioned the wall murals that adorn some of Mexico City's public buildings, murals that integrate revolutionary class politics with new national myths organised around native images and legends. However, Vasconcelos was also the author of a number of studies on racial hybridity. In direct opposition to European notions of race purity and the 'excessive esteem for European and Anglo-Saxon culture' that characterised the pre-revolutionary regime of Porfirio Díaz (Beltrán, cited by Knight, 1990, p.80), Vasconcelos drew on established traditions of race mixing in Latin America to propose that hybridisation was the only way forward for humankind. Thus Vasconcelos identified European racism not simply as an imperialist ideology but one that was subversive of the attempt to develop a new and better form of civilisation in Latin America.

> The British preach natural selection, with the tacit conclusion that world domination belongs, by natural and divine law, to the dolichocephalous man from the Isles and his descendants. But this science, which invaded us with the artefacts of conquering commerce, is fought as all imperialism is fought: by confronting it with a superior science, and with a broader and more vigorous civilisation. (Vasconcelos, 1997, pp.33–4)

Vasconcelos's critique of European racial hierarchies was highly influential when first published in *La raza cósmica* (*The Cosmic Race*) in 1925 (1997). His advocacy of the dynamic and virile qualities of the 'cosmic race', offered revolutionary Mexico a new national myth of identity, one that, at face value, appeared to promise that modernisation and development could proceed apace without the blessing or intervention of Europeans (see Knight, 1990). However, mestizaje has rarely been able to extricate itself from white supremacism, and Vasconcelos's version was, in this respect at least, no different from many others. For although Vasconcelos often appears convinced that when the various racial elements were blended together 'whiteness', 'blackness' and so on would disappear, and racial utopia would be achieved, he also considered that some racial elements were more desirable than others. Thus he argued that the input of 'inferior races' into the mixing process should be both limited and controlled. Indeed, there exists a telling slippage in Vasconcelos's work between the notions of 'mixture' and 'absorption'. As the passage below reveals, the latter process does not promise the destruction but rather the final victory of white racism.

> The lower types of the species will be absorbed by the superior type. In this manner, for example, the Black would be redeemed, and step by step, by voluntary extinction, the uglier stocks will give way to the more handsome. Inferior races, upon being educated, would become less prolific, and the better specimens would go on ascending a scale of ethnic improvement, whose maximum type is not precisely the White, but that new race to which the White himself will have to aspire with the object of conquering the synthesis. (p.32)

Vasconcelos's position provides a stark example of the way the celebration of mestizaje, or race mixing, can both challenge and reaffirm racist categories. Echoes of this aspect of his work may be found in many examples of the nation-building process in Latin America. Within Colombia, for example, mestizaje has simultaneously been associated with racial equality and with Eurocentric discrimination, as explained by de Friedemann and Arocha (1995) in their survey of Colombia's racial history.

> The essence of Colombia's 1886 constitution was an 'either/or' proposition in which 'people' equalled mestizos, and 'not people' equalled indigenous people. The constitution glorified as the goal of progress the conversion of Colombians into a single 'race', speaking one language and believing in a single God ... In 1922 the ideology of whitening (blanqueamiento) as a necessary condition of national progress was used to ... [legally] for[bid] the immigration of people who would be 'inconvenient' for the nation and the development of the Colombian 'race'. (p.65)

The policy and ideology of 'whitening' provide some of the clearest testimonies to the association between whiteness, development and modernity. Immigration policies oriented towards enabling white immigration and discouraging non-white immigration were once commonplace across North, Central and South America. I shall not attempt to survey the many particular histories pertinent to this history. Instead I offer two specific cases, each from a putative 'racial democracy', namely Venezuela and Brazil.

Whitening in Venezuela and Brazil

Both Venezuela and Brazil are sometimes offered as exemplars of nonracist societies. In contrast to the USA, both countries can, indeed, appear very much like the racial democracies their leaders have at various times proclaimed them to be. It is true that in both nations elite groups tend to be light-skinned and the poor dark-skinned. It is also true that almost every indicator of social and economic status shows that, on average, 'whites' do better than other groups in both countries. However, it is equally the case that to an extent greater than in some other countries on the continent (both north and south), there remains considerable mobility between racial–economic strata, and that the correlation between status, wealth and race is more a rule of thumb, a roughly defined tendency, than an iron law. It should also be pointed out that explicit articulations of racial grievances are relatively rare in both countries. Indeed, to the evident chagrin of contemporary anti-racist anthropologists from the United States, the kind of black consciousness politics evident in the latter country remains a marginal current (Twine, 1998).

The adage 'money whiteness' has considerable currency in both Venezuela and Brazil. Of course, the corollary of the notion that money whitens is that poverty darkens. More insidious still is the closely related assertion that respect for whites is based not on racial sentiment but on an admiration of people who are successful, and that prejudice against non-whites simply reflects a revulsion against poverty. In *Café con leche*, Winthrop Wright (1990) has dissected the convoluted connections between economic and racial status evident within Venezuela. Looking back over the country's history he notes that, 'occasionally',

> financial and political success socially whitened black Venezuelans ... For them and their white counterparts, clothes, education, language, social position, and the accumulation of wealth combined to make an individual whiter in the social context. In such a setting the term *blanquear* (to whiten or bleach) had tremendous social significance. (p.6)

Wright continues that the 'emphasis placed on whitening as a prerequisite for social and political mobility suggests the antithesis: black characterised backwardness, ignorance, poverty and failure.' This chain of association did not affect only the lives of European- and African-heritage Venezuelans. All groups were positioned relative to the ideal of whiteness, including native people and other immigrant groups. The link made between backwardness and dark skin colour meant that the path towards modernity, towards national advancement, has often been understood in Venezuela as a route towards whiteness. At the beginning of the twentieth century, the Venezuelan politician Cristóbal Mendoza (cited by Wright, 1990, p.77) explained:

> This is the only sure and easy road we have towards civilisation: to educate, to instruct, but not in the stagnant atmosphere in which we have vegetated during centuries, but rather in the powerful ambience of European civilisation.

It is a familiar plea. To be modern, to be forward-looking, demands a break with a non-European past and an emersion into the new ways and attitudes of 'European civilisation'. The range and scale of the attempt to 'Europeanise' Venezuela and Brazil was considerable. To be culturally sophisticated, to adopt modern economic roles and practices, to build one's cities, roads, farms and factories and arrange one's bureaucracies in an efficient manner: all these keys to advancement were symbolised by things European. And the central and cohering symbol of 'things European' was, of course, the flesh of the European. Thus the practices and ideologies of modernity were *embodied* in the figure of the European, in the person of the white who offers progress as a universally obtainable project, a distributable gift, but whose own body is its defining symbol.

The conundrum of modernity's symbolic reliance on something as apparently fixed and limited as 'white skin' was exacerbated in Latin America by the intellectual impact of Social Darwinism and eugenics. Although the notion that whites represented the most advanced race was current before the influence of these forms of sociobiology, they further codified and naturalised white supremacy in racial discourse. More specifically, Social Darwinism and eugenics were incorporated into emerging narratives of national identity. The conquest and/or assimilation of other races by whites was increasingly deemed not only justifiable but a reflection of a basic law of nature, an inevitable part of national growth. Thus the presence and history of people with significant native or African ancestry was increasingly deemed to belong to a 'dark past', a past which the nation had to struggle against, a past to be contrasted with the white (or, at least, whiter) future. As this implies, many political leaders in both Brazil and Venezuela in the first few decades of the twentieth century were acutely aware, not merely of the undesirability of non-whites in their population, but of a racial contrast between the past and future, between a black and brown pre-modern, primitive and unsuccessful country and a European-heritage, successful country waiting to be born. 'We are two steps from the jungle because of our blacks and Indians,' explained the Venezuelan politician Blanco Fombona in 1912, 'a great part of our country is mulatto, mestizo, and zambo, with all the defects which Spencer recognised in hybridism; we must transfer regenerating blood into their veins' (cited in Wright, 1990, p.72).

Before explaining how the desire to 'whiten' Brazil and Venezuela translated into immigration policy, it may be useful to touch briefly on another aspect of this process, one that is often overlooked: the national geography of whiteness. As I have already hinted, the racialisation of modernity was also a spatialisation. 'European' is, after all, an identity premised on the existence of a racial homeland, Europe. In both Venezuela and Brazil, geographical distance from the centres of the Europeanising process (such as metropolitan areas), was used to map the presence or absence of backwardness, to chart which regions, which places, were part of the preceding era and which were part of the future. In both countries, the forested interiors, often identified with native settlement, as well as those areas associated with African-heritage people, were and are typically understood as distant, exotic, 'awaiting development' or, indeed, 'protection'. Such areas have long been conceptualised both as problems that need to be solved and as representing the unspoilt, excitingly savage, spirit of the country. In either case – whether they are seen to offer headaches for developers or pleasures for primitivists – they are used to connote the past. In other words, whether considered appalling or alluring they are constructed as spatially and temporally remote, their 'survival' an anachronism. The flip-side of this association is the way urban areas have been connoted as points of concentration for whiteness. Most of the main cities in Brazil and Venezuela have acted as both conduits for and sites of expression of the white ideal. Indeed, as with many other cities in the Americas, they have often been built, or rebuilt, in such a way as to evoke European order, rationality and technological superiority. An example is the planning of Rio de Janeiro in the early twentieth century. The rebuilding of Rio was designed to Europeanise the city, to take it away from the past by cleaning it up, ridding it of diseases and making it, in the words of one contemporaneous commentator, an attractive place for 'foreign blood, foreign brawn and foreign capital' (de Barro, cited in Skidmore, 1974, p.131). As part of the city planners' ambition to eradicate the 'dirty, backward and fetid' past, many of the areas of Afro-Brazilian settlement in the city were demolished, and their inhabitants forced to leave. Once these non-white elements had been expelled, the Exposition of 1908 could be held, a celebration of Brazil's technical and economic achievements and bright, modern future.

As with other countries around the world, immigration policies in Venezuela and Brazil have been shaped by concerns about the existing population profile. For many years the principle concern was an explicitly racial one: there were not enough whites. As I have already indicated, racial questions were also economic and social questions. Politicians in both countries framed the issue of white immigration in terms of the need for modernisation. Hence, to enable white immigration was to enable each country's transition from peripheral backwater to economic centre stage, from impoverished and unimportant place to rich and influential one. '[S]trong [i.e. white] immigration', the Venezuelan intellectual Gil Fortoul (cited in Wright, 1990, p.56) observed in 1895, will bring with it 'an industrial and capitalist system that surely will produce considerable changes in distinct manifestations of the national life.' In both countries a policy of national 'whitening' was pursued by both state and non-state interventions in immigration control. Within Brazil, one of the first indications of the former was a government decree, issued on 28 June 1890, permitting 'free entry by persons healthy and able to work ... except natives of Asia or Africa, who can be admitted only by authorisation of the National congress and in accordance with the stipulated condition.'[1] An example of a less formal, non-state, attempt to achieve the same end may be found in the work of the Society for the Promotion of Immigration. Founded in 1886, the society consisted of a group of wealthy Brazilian landowners who had clubbed together to encourage white immigration. Their ostensible motive lay in the claim that the best agricultural workers were European. The society encouraged Europeans to come to Brazil by paying their passage from Europe to São Paulo and arranging jobs for them when they arrived. Most of those who arrived under this scheme were Italian, but nationals from many other European countries also took advantage of the society's generosity. It is indicative of the close relationship and mutual interest between the landowning and political elite that the Brazilian state took over financial responsibility for this programme until its abolition in 1928.

Further state support for the whitening of Brazil was provided in the Constitution of 1934. Article 121 (section 6) states that 'The entry of immigrants into the national territory will be subject to the restrictions necessary to guarantee the ethnic integration and the physical and legal capacity of the immigrants.' The carefully coded message of 1934 was offered in less cryptic form by the decree-law of 1945, which asserted that entry of immigrants should be based on 'the necessity to preserve and develop, in the ethnic composition of the population, the more desirable characteristics of its European ancestry.'

A similar set of legal and civic initiatives may be found in Venezuela. In 1866, Jacinto Reyeno Pachano, minister of development, had rationalised the government's stance in the following terms:

> if the appropriate immigration, that of the European of good physical and moral condition, the industrious and capable European is a moral and progressive element, bad immigration, that which lacks such virtues, far from producing positive good for us will become a germ of immorality and decadence. (cited by Wright, 1990, p.62)

The attributes of the ideal immigrant were summed up in the newspaper *El Derecho* in 1892:

> If to the vigor of nature one joins the perfection, the superiority of intelligence and the active assidulity of the Caucasian race, which one finds for the most part in Europe, the judgement favours European immigration, a decision that becomes irrevocable since experience has confirmed that that race crossed with the hybrid that composes the general population of Venezuela, perfects ours, producing a sane, intelligent, beautiful and strong generation. (cited by Wright, 1990, p.65)

To expedite the development of 'a sane, intelligent, beautiful, and strong generation' Venezuelan governments subsidised the passage of European migrants (as authorised in immigration laws in 1831, 1837 and 1840). The 1891 immigration law stated that 'Individuals of Asiatic nationality or those from the English and Dutch Antilles [who were assumed to be of African heritage] will not be contracted or accepted as immigrants.'[2] The official ban on non-white immigration into Venezuela was enforced until 1945.

However, the racist rhetoric of certain Venezuelan and Brazilian laws and law-makers should not be used to obscure the complex and contradictory nature of racial attitudes in both countries. For, in marked contrast to the USA, there remained a good deal of hesitancy in both societies about explicitly categorising all those who were not white as inferior. Indeed, some of the measures mentioned, such as the restrictive Brazilian decree of 1890, never made it into law (the 1890 decree was not included in the immigration law of 1907). A widespread unwillingness to appear anti-non-white, to appear to be calling for the extinction or extermination of non-whites, was apparent throughout much of the debate. Discussing Brazilian government policies on national whitening, Thomas Skidmore (1974; see also Skidmore, 1985; 1990) explains that, most deputies like most members of the elite shied away from such overtly racist gestures as an absolute colour bar. They believed in a whiter Brazil and thought they were getting there by a natural (almost miraculous?) process. An overt colour bar smacked of the United States, which remained a constant reminder of what almost all Brazilians considered an inhumane (and eventually self-defeating) solution to the ethnic problem. (p.198)

As the example of Vasconcelos's theory of 'the cosmic race' suggests, the ideology of mestizaje, of race mixing, combines racism and anti-racism. The contradictory, ambivalent qualities of mestizaje make it an alluring political discourse for those seeking to appear, simultaneously, as opponents of white domination and advocates of the benefits of national Europeanisation or whitening. Versions of mestizaje were adopted by politicians in both Venezuela and Brazil to evade accusations of racism whilst sustaining a virulent white supremacism. Thus the practice of whitening through white immigration was portrayed, not as an attempt to sustain a discrete or pure white race but as *feeding in* whiteness to a brown and black pool, as improving the racial stocks of both countries by further mixing them with European blood. Such was the strength of the belief in the superiority, the natural vigour, of Europeans that it was widely assumed that the union of white and non-white would naturally whiten the latter, that whiteness would inevitably gain ascendancy. Thus, racial intermixture was not considered a threat to white racial dominance, as it was in the USA, but as the key site of its survival and ascendancy. Venezuelan President Carlos Gómez explained in 1906 that 'The crossing of our sickly and old-looking race with vigorous and young races is indispensable' (cited by Wright, 1990, p.77). A visitor to Brazil from the USA, Clayton Cooper, expressed the matter in more brutal terms in 1917: an 'attempt is honestly being made here,' he explained, 'to eliminate the blacks and browns by pouring in white blood' (cited by Skidmore, 1974, p.74).

At the First Universal Races Congress, held in London in 1911, the director of the Brazilian Museu Nacional, João Bastista de Lacerda, presented a paper entitled 'The Métis, or Half-Breeds of Brazil'. Lacerda disputed the conventional wisdom of many European racial scientists that racial mixture led to degeneracy. 'Contrary to the opinion of many writers', he told his audience,

> the crossing of the black with the white does not generally produce offspring of an inferior intellectual quality ... children of métis have been found, in the third generation, to present all physical characteristics of the white race ... [some] retain a few traces of their black ancestry through the influence of atavism ... [but] the influence of sexual selection ... tends to neutralize that of atavism, and removes from the descendant of the métis all the characteristic features of the black race ... In virtue of this process of ethnic reduction, it is logical to expect that in the course of another century the métis will have disappeared in Brazil. This will coincide with the parallel extinction of the black race in our midst. (cited by Skidmore, 1974, p.66)

Larcerda's views capture perfectly the logic and morality of the 'racial destiny' supported by many upper- and middle-class Brazilians. They could simultaneously claim that theirs was a society without racial preju-dice – because it was a society of race mixture – whilst maintaining a pro-found white supremacism. The matter was neatly encapsulated in a pamphlet written by Caio de Menezes (cited by Skidmore, 1974, p.131) issued in 1914: 'we enjoy an advantage over the United States – the good fortune to have discarded color prejudice with the result that the Negro himself tends to dissolve in the inexorable whirlwind of the white race.'

For many of the politicians and intellectuals in Venezuela and Brazil active during the years when scientific racism was at its most influential in both countries (i.e., 1850–1940) only the 'inexorable whirlwind' of whiteness could produce national salvation. Whiteness was going to blow away the decrepit, parochial past and enable a successful and vigor-ous country to emerge. It would produce, moreover, a *new* society, a bold social experiment. The ideology of mestizaje does not permit a simple copying of Europe, or Europeanness. It projects, instead, an absorption of whiteness as the foundation of societies that are not merely 'as good' as Europe or North America but potentially better (better, at least in part, by virtue of being less racist and because they had achieved the synthesis of the best parts, the most productive energies, of all races).

Relative to their impact on the history and demographic profile of the region, the ideologies and practices of whitening remain neglected topics within political and economic debate. This marginalisation returns us to the role assigned to the United States in discussions on whiteness and racism. For a great number of Latin Americans the very terms 'white supremacism' and 'racism' *evoke* the United States. The latter country symbolises 'race problems'. By comparison – and it is a comparison often drawn – countries in Latin America (especially Brazil and Venezuela) are seen by many of their own nationals, as well as by foreign observers, as not having a particular 'problem'. The social significance of such compar-isons with the USA is examined in some detail by both Wright and Skid-more (see also Twine, 1998). Wright (1990, p.126) concludes that, 'throughout the present century Venezuelans defined racism in terms of the virulent, hate-filled type of discrimination and segregation found in the United States; they have not considered the subtle forms of racism they practice as discrimination'. Skidmore develops the same argument in the context of Brazil. He also makes a controversial additional observa-tion to the effect that the rise in the 1960s of civil rights and other anti-racist movements, alliances and legislation in the United States (as well as anti-colonial independence movements around the world) has under-mined both the international prestige of white identities and the utility of the comparison with the USA for Brazilians who wish to see their country as a racial democracy.

> [T]he achieving of political independence in Afro-Asia and the civil rights revolution in the United States dramatically underlined the loss in prestige for the archaic European-centred culture whose racist assumptions had first led Brazilians to formulate their 'whitening'

rationale. Brazilians had produced that vision of their racial future because it seemed to reconcile the reality of the multi-racial society with the European–North American model of development they sought to imitate. Now that Europe and North America had politically (as well as scientifically) repudiated racism, and now that non-whiteness had become a source of cultural pride and political power both in Afro-Asia and the United States, Brazilians were left with a badly outdated ideal of their racial future … Brazilian opinion-makers are still living with the intellectual legacy of the compromise their parents and grand- parents made with racist theory. (Skidmore, 1974, pp.214–18)

There are various contentions at work in this passage. First, Skidmore reiterates the notion that the advance of anti-racism in the nation that Brazilians have employed to define their own racial tolerance (i.e., the USA), has subverted that comparison and opened up questions about the reality of Brazilian racial democracy. This point may be supported by reference to the rise of black consciousness organisations in Brazil since the 1970s, as well as an increase in criticism of what is now, indicatively, a clichéd phrase, 'the myth of racial democracy'. Writing in the early 1970s, a time when it could still be imagined that anti-imperialism and 'third worldist' perspectives were, or would be, sweeping all before them, it is understandable that Skidmore should seek to connect this contention to a supposed global decline of whiteness as a racial ideal. However, whichever way we excuse it, it remains a bitter irony that the latter observation exhibits what we may term the fallacy of racial chronology, a fallacy Skidmore's own studies in Brazil had helped expose. For the logic of the idea that non-white Brazilians were 'of the past' and whiteness was 'of the future' reappears, albeit turned on its head, in Skidmore's contention that, in contemporary Brazil, whiteness is an ideology of bygone days, a remnant of an earlier and now discredited racism. It is pertinent to note here that, although it is true to say that the connections between scientific racism and ideologies of development were stronger and certainly more explicit in the first half of the twentieth century, the relationship between white identity and modernity has never been wholly reliant on scientific racism. The range and repertoire of white supremacy far exceeds the terms of biological racism and, as events have shown, is quite capable of surviving the latter's relative demise. This is particularly clear in Latin America, where the idea that there are naturally discrete races with unequal capacities has long been suffused with, or been subordinate to, the contention that, in purely biological terms, all races are equal but that white people are associated with the most sophisticated and civilised mode of cultural, social and economic practice.

I asserted at the beginning of this chapter that the social ideal of whiteness remains prevalent throughout the contemporary world. This does not mean that people all over the globe adhere to the myths of racial science. Indeed, it implies that the relationship between whiteness and ideologies of modernity goes deeper, and is more varied, than scientific racism. It also suggests that this relationship is a mutable one; its form and content vary between different places and at different times. In the next section, these ideas will be illustrated with reference to the time and place of neo-liberalism as explored through popular culture, more specifically the Brazilian media 'megastar' of the 1980s and 1990s, Xuxa.

Whiteness and the symbolic economy of neo-liberalism: the Xuxa Show

To say that whiteness '*still* matters' in Latin America is a statement of fact. Yet it is also a little misleading. For if we limit ourselves to identifying an *inheritance* of white racism, to pointing out the *continued existence of traditions from the past*, we are guilty of offering a very partial reality. Whiteness does not merely haunt contemporary Latin America like some disreputable ghost. It is also part and parcel of today's 'symbolic economy'. This section will elucidate and exemplify this point in more detail by drawing on cultural material from Brazil. However, I want to broach this admittedly complex and controversial issue by citing the Peruvian equality activist Patricia Oliart (1997). Oliart is concerned with the impact of neo-liberalism on racism. To understand her concerns we need to know what neo-liberalism is. Basically, neo-liberalism is a form (indeed some would regard it as a pure form) of capitalism. In contrast to state-managed capitalism, neo-liberalism is identified with the 'freeing-up' of the economy by mechanisms such as privatisation and the eradication of restrictive and protectionist interventions in the economy. In Peru, as throughout much of Latin America, neo-liberalism has emerged over the past twenty years as the dominant economic paradigm. This process has been facilitated by the advocacy and economic sway of global financial institutions. Any 'reaffirmation' of the free market is also, of course, a social project, one that ties economic change to a set of expectations, stereotypes and illusions about the benefits of a cosmopolitan, 'open' society. More specifically, Oliart identifies an association of the internationalisation of economic and media interests with the reinscription and reinvention of the white European as the symbol of modernity, of social progress and physical attractiveness.

> What is happening now is that racism is coming back stronger ... Money counts again now and the way you look, we have lots of gyms, that we never had before, all classes doing aerobics and dying their hair, like blond hair ... it doesn't matter if you're not white, you can look white, you can become white, you can wear nice shoes, you can dye your hair, you can get a great body, and if you don't do that, you have a pony-tail, wear ethnic skirts or whatever, then that's your problem. (Oliart, 1997)

As Oliart's observations indicate, whiteness has come to be associated in Peru not simply with the ideals and norms of the old elite but with a more contemporary phenomenon, namely consumerism. Whiteness is connoted as a lifestyle, symbolically tied to the pleasures of a consumption-led identity (pleasures such as 'freedom' and 'choice'). The figure of the white consumer is not an entirely novel one in Latin America. Indeed, the 'white shopper' has been a normative model in the construction of elite female roles in metropolitan cultures of modernity in the region since the late nineteenth century. However, as the ideology and practices of consumerism have become more globalised and socially extensive, this aspect of white identity has become available to a much wider range of people. Now almost everyone is expected to want to exercise their right to immerse themselves in the market.

Neo-liberalism is associated with the deregulation of previously protected economies, enabling them to become more open to outside capital and influence. It privileges flexible, decentralised, non-state-controlled socioeconomic arrangements at the expense of public, state-owned and non- or anti-capitalist forms of organisation. This process has had the consequence of further encouraging North American, Asian and European business interests to move into Latin America. One of the most visible markers of this expansion is the fact that in the downtown area of many cities in Latin America one now finds the same or similar consumer outlets and company logos as those encountered in the USA. The streets are increasingly cluttered with fast food outlets such as McDonald's (or local versions of it), advertisements for Coca-Cola (or local versions of it), and many other brands recognisable as part of 'international' consumer culture. Western goods and symbols are the most visible symbols of the transition to neo-liberalism. Nevertheless, the racialisation of neo-liberalism is full of subtleties and contradictions. Like other forms of capitalism, neo-liberalism provides arenas of resistance and egalitarian change alongside its socially conservative and anti-egalitarian tendencies. It is interesting to note in this regard that it has become associated with the assertion of inter- and multiculturalism in Latin America (see Laurie and Bonnett, forthcoming). This alliance may seem a strange one but is, in fact, quite logical. Neo-liberalism seeks to create flexible, market-led economies. It needs geographically and socially mobile societies. Hence it requires the dismantling of those barriers that stand in the way of such flexibility. Some of the most significant of these barriers are ethnic and racial prejudices and loyalties. Moreover, there is evidence to suggest that consumer identities can also be anti-racist identities. For example, the deployment of African-American and Latino style and 'attitude' amongst youth groups in Latin America appears to be offering an avenue into both consumerism and the ostensible rejection of whiteness. However, this latter current remains a slender one in most countries in the region. Indeed it is surely of note that the terms 'multiculturalism' and 'anti-racism' and, in certain contexts, 'black' and 'Latino' identities, are often understood in Latin America as imports from the USA.

It is true that neo-liberalism offers possibilities of opposition, especially within the realm of aesthetics and consumer choice. It is also true that it deploys discourses of equality in those circumstances where inequality is seen to stand in the way of social and economic flexibility. However, these tendencies are continuously compromised by the much more powerful normative and anti-egalitarian currents within neo-liberalism. Multiculturalism and anti-racism may, sometimes, be encouraged by free-market reforms, but it is a celebration of equality that takes place within societies where the appeal and power of whiteness is being both sustained and, at least in certain areas, strengthened.

Maria da Graca Meneghel became a 'megastar' in Brazil in the early 1980s. It was also at this time that she adopted the name Xuxa. She is a television personality, a white star in a predominately non-white country. Reflecting a pattern evident throughout Latin America, the Brazilian media and advertising industry routinely portray an image of the country as white. White models, actors, newscasters and other 'personalities' are as ubiquitous as black ones are rare (see Moore, 1988; see also Kottak, 1990). Xuxa is not exceptional because of her whiteness. However, the scale of her personal success is distinctive. She is a media phenomenon, a household name throughout the country, whose life story and achievements have captured the interest of both journalists and scholars of Brazilian popular culture.

The *Xou da Xuxa* (*Xuxa Show*) is Xuxa's principal vehicle, a children's programme that also attracts a large adult following. Although its studio audience and musical guests are overwhelmingly Brazilian, the show is similar to the kind of light entertainment programming now found all over the planet. The musical style is relentless disco pop, the presenters look and sound, if not European, then 'of the West'. The atmosphere is one of boundless 'zany' fun and endless enthusiasm for the latest consumer fashion. Xuxa has been able to extend this format beyond her television show and into numerous other arenas of cultural production. The range of items issued under her name, from songs to shoes, from dolls to films, is enormous. However, her fame is rooted in television, a fact that reflects the role of the TV as the primary medium of consumer and, indeed, national identity in Brazil. Brazilians watch a lot of television (a study in São Paulo showed that 95 per cent watched regularly during the week. Brazilians also have a lot of television sets, ranking fourth in the world in terms of numbers owned: Simpson, 1993). Moreover, television production companies in Brazil are large and influential. The company which produces Xuxa, Globo, owns about one-third of the broadcasting stations in Brazil. Such media giants have been able to shape the style, content and social function of television. According to Amelia Simpson's (1993) study *Xuxa: The Mega-Marketing of Gender, Race, and Modernity,* 'Globo has become known as the instrument largely responsible for creating a national culture in a country with many separate regional identities' (p.63). This televised 'national culture' contains several elements, but at its core is the figure of the consumer. Audiences are consistently bombarded with images of goods for sale and attractive lifestyles and people to emulate. The seeming incongruity of such programming in a country where the majority of the population have only a small disposable income has not been lost on social critics within Brazil. Nico Vink (cited by Simpson, 1993, p.44) suggests that a 'contradictory consumption society has been formed in which, for the majority of the population, the images of consumption goods have increased more than the real access to these goods.' This process relies on the commodity functioning primarily as image, as a pacifying spectacle, an icon of fulfilment and identity always out of reach but always to be reached for.

Product placement on the *Xou da Xuxa* is explicit. For example, in 'The Coca-Cola Game' the contestants wear Coca-Cola logos on their t-shirts whilst rushing around in the attempt to put together a picnic, a meal that consists of hot dogs, drinking straws and, of course, Coca-Cola. The idealisation of 'American-style' consumerism also emerges in the game called 'Yes or No':

> This game ties triumph or defeat to the value of goods and delivers a lesson in the authority of the market place. No particular product is promoted; instead, the issue of relative value is raised. The single participant, sitting in a soundproof booth and unable to see what Xuxa and the audience are viewing, is required to agree or not agree to trade one item for another. Without knowing what the items are, the payer must respond 'yes' or 'no' when a light is switched on. Xuxa might ask, for example, 'Will you trade your video game for a bicycle?' or 'Will you trade your TV for a pacifier?' Afterward, Xuxa releases the player from the booth, awards the prizes, and explains what the boy or girl traded away. This game of chance stimulates in the audience a kind of frenzied appraisal of goods.... 'Yes or No' teaches that a video game is more valuable than a bag of marbles and that exercising one's options, even with limited choices and in compelling circumstances, is a way of defining oneself as a winner or a loser. (Simpson, 1993, p.86)

Yet, although it can sometimes appear as if the real stars of Xuxa's show are the products on display, it is Xuxa's personal advocacy of consumerism and her deployment of whiteness that animates and structures the programme and its spin-offs. As this suggests, the specific visual mechanisms that communicate her appeal also tell us something about the kind of whiteness that she represents. I have already mentioned the theme of pleasure, of white identity being offered as a way of life that is associated with the thrills and freedoms of participation in a market economy. This chain of connotations may be extended further to include excitement and transgression. In the *Xou da Xuxa,* all these attributes are conflated, condensed into one sign, as 'sexiness' ('sexy' is a word now known in its English form in many non-English-language-speaking countries). As this implies, Xuxa's image in Brazil is not simply that of a white person; nor, indeed, is her whiteness seen as merely aesthetically superior. Xuxa offers something more particular: the sexual desirability of the young blonde white women. Indeed, the visual representation of Xuxa, in publicity and poster material, in film spin-offs and in the show itself, is organised around the clichés of soft pornography; more specifically, of the flirtatious sexual availability of the submissive, feminine and childlike blonde. This role is continued in the interactions of Xuxa with the children on her show. The show's child guests are encouraged to engage Xuxa in erotically suggestive banter. Indeed, it may be argued that in the *Xou da Xuxa* the themes of whiteness and consumerism find expression in an infantilised sexuality, in which sexual attraction is synonymous with 'naughtiness' and the pre-pubescent body. Xuxa's blondeness may be seen, then, as a key signifier not simply of her whiteness but of her affiliation with the non-adult and, relatedly, with submissive femininity.

By the mid-1980s, Xuxa had become established as Brazilian television's principal star, a success that was soon to be mirrored in other countries throughout Latin America. Xuxa has also made inroads into the Spanish-speaking television market in the USA. However, to many North American commentators, the most striking thing about Xuxa was the incongruity of a supposedly non-white country such as Brazil producing a white star (Brooke, 1990; Chapoval, 1992). This putative tension provides the central problematic of Simpson's analysis of the Xuxa phenomenon:

> She asserts the superiority of whiteness through her own image and its many manifestations, including the blond imitation Xuxas, the Paquitas, who are the envy of virtually every Brazilian girl at one time or another. Blondness is a norm of attractiveness that is inaccessible to most people in Brazil, the country with the world's second-largest number of people of African descent. Yet Xuxa's representation of all-white aesthetic is symptomatic, not prescriptive. The star's promotion of the white ideal functions only with the complicity of an audience eager to view blond beauty. (Simpson, 1993, p.7)

Simpson is surely right to emphasise the 'inaccessible' nature of Xuxa's whiteness, an attribute which complements the 'out of reach' nature of the consumerist fantasy she embodies. However, whether this process is unique to Brazil, or Latin America, is debatable. Western societies are, after all, also entangled with the endless quest for fulfilment through spectacles of consumption and ideal racialised beauty. Capitalism relies on precisely this type of personal investment, of a constant effort to emulate lifestyles beyond one's reach. If we accept this contention then the representation of Xuxa by Western commentators as an extraordinary phenomenon, indeed, as ludicrous, starts to appear myopic and self-deceptive. Indeed, such portrayals begin to appear like an attempt to present Brazil as the comfortably distant location for the racialised dramas that characterise all contemporary forms of capitalism. Emphasising the foreignness, the oddity, of Xuxa has a dual, if contradictory, comfort value for European and North American critics and their readers. It reassures them that non-Western societies are dominated by 'our' norms – that 'they' want to become 'like us' – whilst at the same time implying that the West has no investment in the racialisation of consumerism. One is reminded of Henry Champley's condescending glee, introduced at the start of this chapter, in observing what he took to be non-white people's constantly unsuccessful attempts to emulate white role models. Yet who could, or can, ever live up to such archetypes? Fantasies of extraordinary beauty and fabulous wealth are no more real, no less deceptive, in the West than anywhere else.

The Xuxa phenomenon may be taken as an example of the power of discourses of white supremacy within contemporary Brazil. Yet this process cannot be understood merely in terms of the imposition of a racist ideology on Brazil. Rather, Xuxa has been actively produced within Brazilian society. Her symbolic repertoire reflects less some pre-formed 'white identity' that has been manufactured in the West and exported around the world than a process of selection carried out in Brazil itself, a process that has created a version of whiteness for the Brazilian market. Pursuing this theme, it is pertinent to note that the sub-pornographic infantilism of Xuxa has many of the hallmarks of kitsch, a popular aesthetic mode that exaggerates and complicitly parodies existing cultural traditions. Thus we may be tempted to argue that Xuxa *appropriates* whiteness, even that she represents a kind of resistance to more conventional and reverential attitudes towards this identity. Certainly, she makes whiteness look rather silly, an empty-headed and immature search for short-term pleasures. However, although it is tempting to locate transgression within non-Western engagements with whiteness, it seems clear that the Xuxa story is much more about domination than it is about resistance.[3] It may be claimed that Xuxa translates, mutates and otherwise disrupts the conventions of whiteness. In part it is true. But it is a minor truth. The more prosaic and historically familiar aspects of the Xuxa phenomenon – namely its white supremacism, its Eurocentrism, its association of whiteness with modernity – are also the ones that tell us most about Brazilian society and her place within it.

Deploying and adapting whiteness: imagining self and other in Japan

Over the past one hundred years or so, white identity has been used to construct a series of competing and overlapping categories in (and such as) 'Asia' and the 'Third World'. Often this process has involved these identities being defined as *not* white/Western or as *against* whiteness/Westernness. Many societies, though, are difficult to place within such neat dualisms. Japan provides us with an example of a country that can often appear both Western and non-Western, aligned to 'the white club of nations' but not – or so it seems to many – a natural member of that club.

It is pertinent to recall that some Japanese politicians in the early and mid-twentieth century sought to depict their country as the leader of the 'coloured' or 'non-white' people of the world against the West. This sentiment was one of the factors that lay behind the Japanese government's attempt to include a clause on racial equality in the League of Nations Charter, adopted in 1920 at Versailles (a more specific motivation was to attack those racist laws in existence in the USA that were preventing Japanese immigration to that country). Although it was blocked by the USA, Australia and Britain, Japan's diplomatic effort established the country's ability and willingness to act as a voice for non-white peoples (Naoko, 1989). Indeed, in 1935 the African-American intellectual W.E.B. Du Bois (cited by Füredi, 1998, p.44) noted that 'Japan is regarded by all the coloured peoples as their logical leader, as the one non-white nation which has escaped forever the dominance and exploitation of the white world.' Throughout the Second World War, the Japanese government sought to capitalise on such aspirations; there existing what Wagatsuma calls a 'hidden sub-current' (1967, p.139) within Japanese propaganda which positioned Japan 'as the "the champion of the colored nations", [fighting] against the "whites"'.[4]

However, by the time the notion of the 'Third World' emerged in the 1950s,[5] the Japanese economy and Japanese culture were so thoroughly enmeshed in the ambitions and networks of 'Western capitalism' that this earlier non- or anti-Western collective identity appeared increasingly irrelevant. Indeed, Japan's presence at the Bandung Conference (a summit of leading politicians from 'non-aligned' African and Asian nations held in Bandung, Indonesia, in April 1955 and commonly regarded as a founding moment in the forging of an anti-colonial, Third World identity), was greeted with suspicion by many other delegates (see Ampiah, 1995). The fact that Japan was attending the conference at the request of the United States (it was felt by US diplomats that Japan's attendance might weaken the presence of communist China), and that some East Asian nations at Bandung had had recent experience of Japan as a colonial power, heightened the sense that Japan was there in the role of outside observer rather than as an active participant. In the late 1950s and the 1960s the perception that Japan could no longer position itself as an authentic non-Western society was strengthened by the ability of Chinese politicians to take on the mantle of leaders of the Third World. This manoeuvre had the effect of further cementing the association of capitalist, Western and white identities. An extreme version of Chinese 'Third Worldism' was recorded by Louis Barcata (1968, p.194) in his study of Chinese communist intellectuals derived from interviews conducted in 1967. These ideologues, he noted, foresaw 'an epic struggle between the races – an Armageddon in which China would lead the exploited colored peoples in their battle against the powers of white reaction.' As Dikötter (1992, p.195) further records, such '[r]acial hatred reached a peak during the Third Afro-Asian Solidarity Conference in February 1963, when the Chinese delegates vehemently insisted that the 'white' Russians would never commit themselves whole-heartedly to the anti-imperialist struggle.'

In the post-Cold War era, 'Third Worldism' has appeared increasingly susceptible to fragmentation and political redundancy. By contrast, the notion of 'Asia', and more particularly of 'Asian values', seems to have been fortified. 'Asian values' have been cited as the defining features and key to the success of the so-called 'tiger economies' and other emergent economies in East Asia. First propagated by the government of Singapore in the early 1990s, the notion of 'Asian values' has been taken up by a variety of regimes in the region, often in an attempt to secure their own legitimacy. The 'values' considered 'Asian' tend to be conservative, traditionalist, solidatitistic (a value that contains both egalitarianism and nationalism) and work ethic-oriented. This collection of ideals is often understood as rooted in Confucianism and/or Islam. Perhaps even more importantly, they are values considered either not present in the West or as having been lost by the contemporary West. Hence, the latter is cast as materialist, decadent, individualistic and alienated.[6]

Japan has tended to be treated as a special case in East Asia. In part, this may be said to be justified by Japanese leaders' traditional reluctance to see the country as part of a wider 'Asian community' of nations. Indeed, rather than seeking to define Japan against the West in order to align it with Asia, Japanese intellectuals and politicians have often appeared to claim Westernness in order to differentiate Japan from Asia (the notable exceptions to this phenomenon being associated with periods of ultra-nationalism, most notably before and during the Second World War). In late nineteenth- and early twentieth-century Japan, as in China, indige-nous traditions of valuing white skin colour became integrated with Western categories and assumptions concerning race, more specifically the superiority of Europeans (Leupp, 1995). Positive attitudes towards whiteness thus became mixed up with positive attitudes towards Europe-ans as 'the white race'. Such ideas persist in contemporary Japan. Hiroshi Wagatsuma's (1968) study of Japanese skin colour preference found a persistent tendency to ascribe beauty and superiority to 'the white race' (see also Kitahara 1987). Wagatsuma argued that

> Japanese eyes, despite cases of plastic surgery, may keep their Oriental look, but through these eyes Japanese see themselves as part of the modern Western world conceptualised in Western terms. (p.435)

Wagatsuma's conclusion is a familiar one. Histories of Japan have long seen the themes of Westernisation and modernisation as inextricably linked and central problematics of that society's development over the last 150 years. Particular emphasis has been placed on the way Japan has come to define itself in relation to the West, and the concomitant dimin-ishment of the established tradition of viewing Japan in terms of its asso-ciation with China (Sato, 1997). This way of understanding the country's development posits the Meiji period (1867–1912) as the founding era of modern Japan. Sometimes termed the 'Japanese Enlightenment' (for example, Blacker, 1969), the Meiji period saw Japan 'open up' to the West (in contrast to the previous two centuries of isolation). To provide a spe-cific example of this process we may turn to the writings of the Japanese intellectual often considered one of the most important exponents of 'enlightenment' and 'Westernisation', Fukuzawa Yukichi (1835–1901). Fukuzawa was an ardent advocate of Western approaches to a wide range of social and technical practices. Indeed, he identified Westernisa-tion with civilisation.

> We cannot wait for our neighbour countries to become so civilised that all may combine together to make Asia progress. We must rather break out of formation and behave in the same way as the civilised countries of the West are doing ... We would do better to treat China and Korea in the same way as do the western nations. (cited by Blacker, 1969, p.136)

The employment of the West as the benchmark of civilisation required the displacement of the influence of China and the absorption of Western social habits and technologies. However, this process of absorption had a somewhat contradictory intent. Its principal aim was not to achieve assimilation into the West but Japanese independence. The latter, Fukuzawa argued, could only be secured with 'modern', 'rational' – i.e.,

Western – approaches to warfare and political governance: 'The only means of thus preserving our independence is to adopt western civilisation.' 'A country which abides by reason,' he added, 'cannot be assailed from without' (cited by Blacker, 1969, p.68).

Japan's turn towards the West led to the Japanese word for foreigner, *gaijin* (literally 'outside person'), becoming virtually synonymous with 'white person'.[7] However, the cultural emulation of white people could carry subversive connotations. In a commentary on Japanese cultural hybridity, Sakamoto (1996) concurs that Fukuzawa's concern was to adopt Western values only in as far as they could be used to strengthen Japanese national ambitions, and to secure its superiority over other societies in Asia. However, he makes an additional point:

> [that] the transplantation of 'Western' civilisation to Japan is possible at all disturbs the West's identity as the only subject of history as civilisation, as the only author of the narrative of world history. Thus, although reinscribing and grafting the jargons of civilisation into the Japanese context, Fukuzawa created a difference that challenged Western Orientalist domination. His discourse positions Japan within the dominant narrative of Western modernity, without being totally overwhelmed. (p.121)

In fact, the affirmation and rejection of Western values have often proceeded simultaneously in Japan. More specifically, the assertion of a fundamental difference between Japanese and Western approaches has been used to locate the *unique* nature of Japanese identity and Japanese modernity. Peter Dale's (1988) analysis of *nihonjinron* (literally, 'discussions of the Japanese') literature – a huge array of popular and scholarly texts that have sought to define the Japaneseness of Japan – locates a persistent dualism (in which the West and Japan are established as opposites) existing alongside the translation of Western values into Japanese terms (for example, the values of capitalism, consumerism, entrepreneurialism). An example cited by Dale is Tsunoda Tadanobu's best-selling book on the 'Japanese Brain' (published in 1978). Following the pattern established in other texts within the *nihonjinron* tradition, Tsunoda offers a vision of the Japanese as a unique people whose mental processes evidence a distinctive holism, a characteristic evidenced by comparison with the more fragmented, alienated nature of the 'Western mentality'. As summarised by Dale (pp.189–90), Tsunoda asserts that amongst the Japanese the left brain 'unifies the perception of the natural acoustic phenomenon (linguistic, musical and natural sounds), whereas these are treated separately … in the Western brain.'

The tendency within *nihonjinron* literature to pose individualism and materialism as 'Western values' has also been argued to have a political function in Japan, namely the concealment of these tendencies in Japanese capitalism. Describing this process as 'cultural exorcism' (p.40), Dale claims that

> repudiation of Western conceptual models stems from the fact that these latter tend to expose, by force of analogy with earlier Western life,

aspects of tenacious medievalism operant in the Japanese version of capitalism. In this sense, then, the nihonjinron's picture of Japan's alternative modernism disguises both a sympathy for feudal values and critical hostility or revolt against the logic and institutions of bourgeois culture. (1988, pp.44–5)

The expansion and relative success of the Japanese economy since the 1950s has meant that it is now a major power *within* Western politics. Japan has not only been admitted into Western financial and political mechanisms but has become a core component of the world's dominant political current and power block, namely 'Western capitalism'. This process has been accompanied by the 'Americanisation' of post-war Japanese culture (see Tsurumi, 1987) and, at a more fundamental level, the realignment of notions of 'the West' in Japanese discourses of self and nation. Western identity has become increasingly deployed, become ever more familiar, as part of the set of things that defines what it is to be Japanese. Within this process the whiteness, the Europeanness, of Westernness emerges as a disruptive force, capable of creating moments of tension, of disjuncture, within the country's allegiance to the West. The sense that Japan is 'not really' Western, that it is 'pretending', that it exists within the club of Western powers as a stranger, can erupt into even the most official of forums. 'Looking at television's coverage of national leaders on such occasions as the summits of the industrial democracies,' notes the director of the International Research Centre for Japanese Studies in Kyoto, Hayao Kawai (1998, p.144), 'one can see at a glance that the Japanese prime minister, although dressed like everybody else, is different from the rest.' The sense that Japan 'stands out' in the company of Western nations is tied by Hayao to the Japanese people's uniquely intense interest in questions concerning their own national identity. 'The Japanese alone, it seems,' Hayao writes,

> are obsessed by questions about who they are and what their culture is about. In a sense, this is only natural ... among the countries known as advanced nations, only Japan lacks a Judaeo-Christian background and has a yellow race. (pp.143–4)

The tension within Japan's entry into 'the white club' of Western nations can, in part, be glossed over by the replacement of the word 'Western' with other terms, such as 'advanced' or 'developed'. Yet this manoeuvre also has the effect of evoking the racialised nature of these categories. What, after all, does it mean to be 'advanced', or 'developed', other than to be aligned with European modernity and civilisation? Despite the success of the Japanese economy these associations have yet to be severed. Thus the whiteness of the West, the fact that the one connotes the other, continues to inform the historical and geographical imaginations of the Japanese. As this implies, the figure of the white person must be placed particularly carefully in Japanese discourses of self and other. The white must be simultaneously 'like us' and 'not like us', both foreign and not foreign. Perhaps a suggestive rephrasing of this problematic would be to say that whiteness is deployed in Japan as an otherness that is owned, an alterity that is claimed and assimilated. Drawing on the research of the anthropologist Millie Creighton (1995; 1997), I will explore this seemingly contradictory dynamic further by turning to the figure of the white in Japanese advertising.

The ambivalences generated when non-white national and ethnic identities are constituted through discourses of modernity are not, of course, unique to Japan. In the discussion of Latin America provided earlier in this chapter it was shown how whites are used as symbols both of economic and social conservatism and of a consumerism connoted with pleasure and transgression. The media's use of the 'beautiful', 'sexy' white is just as apparent in Japan. A creative director at one advertising agency provided Creighton (1997, p.216) with the following explanation 'for the abundance of gaijin' in advertising images,

> for a long part of Japanese history, from Meiji at least, we have always been looking at Western countries as progressive ones. These were places that Japan had to catch up with. From this there developed a sort of complex – 'it's a white world'.

White Westerners in Japanese advertising tend to be used in roles considered unacceptably daring for Japanese models and/or as symbols of consumerist excitement. Thus, for example, whilst naked and semi-naked *gaijin* women have been relatively common in poster and television campaigns, the same cannot be said of Japanese women. As explained by another advertising employee (cited by Creighton, 1995, p.145), 'ads can't use Japanese women for such nude scenes because it is too realistic, so gaijin are used.' As Creighton (1995, p.137) explains,

> Gaijin are much more likely to be shown overtly breaking the conventional rules of Japanese society, or as individuals who struggle incompetently with the habits and customs of Japanese life. Nude representation of gaijin, particularly naked shots of the upper bodies of both men and women, are common in advertisements for products and services where naked depictions of Japanese would be inappropriate.

Creighton goes on to explain that advertising campaigns that wish to sell a particular consumer product on the basis of its appeal to possessive individualism (i.e., as items of personal satisfaction, to be selfishly guarded), find it appropriate to employ the bodies and the language of white Westerners:

> Foreigners often provide a safer mechanism for expressing selfish sentiments in a culture which has long frowned on wagamama, or self-centred concerns. Related to this is the prevalent use of the English word 'my' in advertising and product names rather than the Japanese equivalents watakushi no, watashi no, or boku no. Ads and product labels commonly refer to 'my jeans', 'my car', 'my home', 'my peanut butter' and even – as I saw in a small town in Shikoku – 'my toilet paper'. (1995, p.146)

The deployment of the English word 'my' is suggestive of the way a particular attribute of capitalism (in this case its possessive individualism) can be sustained in Japanese society by being 'othered' at a symbolic level. A kind of knowing game is enacted that allows the pleasures of selfish consumerism to become all the more tantalising by being cast as, simultaneously, foreign, transgressive and entirely available. This process finds echoes in many societies around the world, but the strength of the Japanese economy, its place – albeit an ambivalent one – within the West, appears to accentuate it. Indeed, it may be suggested that the *gaijin*, his/her body and his/her rhetoric of selfishness are less likely to be read as an imposition from outside than as reflective of Japanese agency, as products of Japanese prioities, a Japanese manipulation of the identities available within capitalism.

The theme of Japanese possession of whiteness has been directly addressed by Creighton (1997). The illustration she chooses to introduce the issue is a Japanese cartoon of an encounter between a Japanese woman and African-Americans. In the cartoon, the African-Americans are aggressively remonstrating with the women, presumably because her clothes feature prints of 'Sambo' figures. The Japanese woman is represented 'with stereotypical white features and hair colouring' (p.228). The US context of this image may be significant: the alignment of Japanese-Americans with and as whites in the USA is an emergent tendency noted by Hacker (1992; though see Lipsitz, 1998). However, Creighton uses the cartoon to make a broader point, noting a shift in Japanese culture towards the occupation of whiteness as a site of dominance within the global economy. The 'Japanese have entered the symbolic space of "white",' she asserts, 'a space suggesting privilege, economic and political prominence' (p.228). If we accept this argument then the advertising cliché of the sexualised, transgressive *gaijin* becomes even more intimate, and exciting; she or he becomes part of a process of symbolic transposition that is *understood* as a symbolic transposition, a process where whites are being employed to represent elements of a *shared* (white?) culture.

What happens, though, when being Japanese appears as not simply as good as being white but better? Within Western advertising in the late twentieth century, Japanese products began to be sold on the basis of their association with the technologically and aesthetic superiority of Japanese society (Moeran, 1996). The notion that Japanese society, especially its technological and business practices, may, in fact, be superior also began to find expression in Japanese advertising. The white began to be cast as someone unable to keep up with the Japanese, as someone who was not capable of emulating Japanese efficiency and standards. Again we may turn to Creighton for a depiction of this process:

> If for decades advertisements reflected a 'gaijin complex' that 'it's a white world', with Japan's reclaimed assurance in its own cultural identity, gaijin faces are now used to suggest that maybe it is, or should be, a Japanese world after all. Humour in recent commercials is provided by beautiful and elegantly attired gaijin women trying to tell jokes in Japanese but stumbling inadequately with the language. A commercial for the Osaka Keikin shopping mall features a gaijin repeating the phrase, 'I can't keep up with the Japanese'. (1995, p.149)

Another example of this phenomenon is found in an advertisement from the early 1980s for the National Rice Council:

> According to this advertisement it is no longer the Japanese trying to 'catch up' with the West. Instead a white businessman in Japan is trying to determine why the Japanese are the front runners. The white businessman sits, holding a bowl of rice in his hands, and says to himself, 'I wonder what makes Japanese business so successful. It must be the rice they eat'. (Creighton, 1997, p.220)

In a society where (Japanese short-grain) rice is connoted with national identity and purity, the white businessman's dilemma is particularly challenging: it is as if he can neither grasp nor hope to connect with the organic, natural essence of Japaneseness. Such examples suggest that the failed and exhausted white may have begun to provide a new and distinctive figure through which to define Japanese identity.

The partial dethroning of European-heritage people as representatives of a superior 'white race' does not necessarily imply the abandonment of whiteness as an ideal or model in Japan. As I have already implied, racial whiteness appears to be available for assimilation, for adaptation and adoption within Japanese society. I would tie this admittedly speculative suggestion with another one. For it strikes me that 'traditional' notions of whiteness in Japanese society – that is, whiteness as a valued, non-racial marker of purity, beauty and superiority – may be able to be redeployed in order to envision a renewed Japanese claim on whiteness. It is interesting to note in this regard that Wagatsuma's interviews with Japanese men and women revealed a distinctly subversive current, namely an attempt to question the whiteness of Europeans. The ugliness of European whiteness as compared with Japanese whiteness was mentioned by several of his informants. More specifically, it was argued that *European-heritage people do not possess white skin but transparent skin*. Three respondents' views are cited below:

> This may be completely unscientific but I feel that when I look at the skin of a Japanese woman I see the whiteness of her skin. When I observe Caucasian skin, what I see is the whiteness of the fat underneath the skin, not the whiteness of the skin itself. Therefore, sometimes I see redness of blood under the transparent skin instead of white fat. Then it doesn't appear white but red.

> I have seen Caucasians closely only a few times but my impression is that their skin is very thin, almost transparent, while our skin is thicker and more resilient.

> The Caucasian skin is something like the surface of a pork sausage, while the skin of a Japanese resembles the outside of kamaboko [a white, spongy fish cake]. (cited by Wagatsuma, 1968, pp.142–3)

'This may be completely unscientific' notes the first interviewee – but in a world where race is no longer regarded as a matter of science, does the European claim on whiteness still have to be respected in such terms? Is whiteness becoming open to deracialisation, to being absorbed (back?)

into symbolic repertoires that no longer privilege any particular race as its rightful owner? Well, perhaps. These questions are worth posing, but on the evidence provided in the rest of this book I would have to admit that even an affirmative response would need to be followed up by a swift, 'but not much'. The racialised nature of the central conceits of our era – modernity, civilisation, progress and their various offshoots – mean that detaching whiteness from the figure of the European still appears counterintuitive, almost unthinkable.

Conclusions

In some areas of India, the Hindu goddess Durga is traditionally depicted astride a tiger. And, traditionally, she is painted black or brown. However, recent years have seen her colour change. She has become paler, whiter. A BBC reporter investigating the phenomenon was told that she 'used to be brown but in this day of TV people want more pleasant images … Indian men today prefer fairer women' (BBC, 1998). Durga's change of appearance is, in itself, a minor story. Yet it reflects a much wider process. Despite the supposed death of scientific racism, the idealisation of whiteness continues apace. I have argued in this chapter that this process is bound up with modernisation and that one cannot understand the latter – surely the core social and economic force of our era – without appreciating its dependence upon its naturalisation and embodiment in racial whiteness.

The triumph of European whiteness has been to transform itself not simply into a norm of and for bourgeois Europeans but for all humanity. Of course, most people's lives are affected by a variety of ethnic and racial norms. Whiteness is not necessarily dominant in each and every racialised situation. Nor, where its presence is felt, is it usually able to be clearly separated from other identities. However, there can be few people left in the world who do not employ whiteness to define at least a part of what they are. Its presence is particularly powerful in the construction of pannational identities, such as 'Asian', 'Third World', 'Black' and so on. But many, more local, identities also make use of whiteness, as something to aspire to, to define oneself against or, as in Japan, in a more complex and tense relationship of assimilation and refusal.

White identities have been deployed in the dissemination and maintenance of power across the globe. Sometimes, this power has been crudely and explicitly racial – the whites conquering the non-whites – but more often, and increasingly, the symbolic repertoire of whiteness has been more subtle. Whiteness has become less a sign of fear, of military conquest, something that is theirs and that we cannot share in, and more one of the defining symbols of modernisation, a process that we are all encouraged to identify with or define our ethnic uniqueness against. To say whiteness is now *less* a sign of fear does not mean that its relationship to military power no longer matters. The ability of the country understood by many around the world to be the white power *par excellence*, the

USA, to intervene militarily around the world has, if anything, increased since the end of the Cold War (for discussion of the contemporary relationship between 'whiteness and war', see Lipsitz, 1998). Nevertheless, with the diminishment of direct colonialism and the emergence of an ever more globalised capitalist world system, the preponderant tendency appears to be towards a situation where whiteness is *not so easily recognised as an exterior threat, as an enemy from without. Increasingly, it is something that exists within popular culture as a symbol of 'our progress' and 'our pleasure'.* This is not to argue, that whiteness is today a free-floating signifier, part of some endless moment of creative play to be constructed or disposed of in any way anyone, or any particular society, sees fit. The power of whiteness continues to be generated by its relationship with social and economic hegemony. As with many of the most succesful forms of oppression, whiteness has been internalised, not merely as a sense of inferiority but as a symbol of freedom, of excitement, of the possibilities that life can offer.

1. All translations of Brazilian decrees and legislation are derived from Skidmore (1974).

2. All translations of Venezuelan decrees and legislation are derived from Wright (1990).

3. A similar point must also be made in response to many less obviously conservative cultural practices in Latin America, practices that contain moments or themes that appear to be racially counter-hegemonic. The racial politics of carnival is a case in point. For example, in his account of the dominance of white skin as the norm in Nicaragua Lancaster (1991) notes the existence of 'moments of reprieve, flashes of rebellion' (p.347):

> in carnivalesque festivities like Santo Domingo, negritude is symbolically elevated over whiteness, and things Indian supersede things Spanish. Celebrants appear drenched in grease and wearing Indian costumes. Youths blacken their own and their elders' faces. As though in revenge against white envy and color climbing, blackness thus assaults the entire community, triumphantly asserting itself as a reversal of the people's baptism into a Spanish Catholic Church (p.347).

Yet as Lancaster notes 'to date, this spirit of rebellion has not much escaped the confines of carnival'. Such 'moments' are symbolised as precisely that, brief temporally and spatially restricted sites of fantasy which by celebrating the exotic and transgressive qualities of non-whiteness confirm the centrality of whiteness. They may speak of a desire to resist, but this quest is articulated in the language of white supremacy.

4. The alliance of the Japanese with other non-whites also enabled the establishment of several pan-racial groups in the USA. The Colored People's Union, established in Seattle in 1921 by Japanese immigrants and African-American followers of Marcus Garvey, was open to all 'except the white or Teutonic races' (Taylor, 1991, p.426). Another group, the Pacific Movement of the Eastern World, established in 1932 in Chicago, sought to bring together all the non-whites of the world under the leadership of Japan. Lipsitz's (1998) analysis of these forms of activism stresses how seriously they were taken by the US government. He cites one Department of Justice report from the 1920s warning that 'It is the determined purpose of Japan to amalgamate the entire colored races of the world against the Nordic or white race, with Japan at the head of the coalition, for the purpose of wrestling away the supremacy of the white race and placing such supremacy in the colored peoples under the domination of Japan' (p.193).

5. First used by the French demographer Alfred Sauvy in 1952, the expression 'Third World' was quickly taken up by the leaders of ex-colonial nations (especially Nehru in

India, Nasser in Egypt and Sukarno in Indonesia), who used it to express the formation of a 'non-aligned' grouping, beholden to neither of the two superpowers (Worsley, 1964; 1984).

6. However, precisely because it is a relatively recent construct, it is difficult to gauge the adaptability and durability of the notion of 'Asian values'. The crisis within the economies of East Asia in the late 1990s may be said to have offered a test to its central conceit, although at the time of writing, it is premature to judge the form and impact of this challenge. It certainly appears as if some regional leaders are beginning to find the term too inclusive to provide effective support for their own particular brand of politics. Writing in 1999, Bell reports (p.6) that

> Singapore's Senior Minister Lee Kuan Yew … has recently abandoned the use of the term 'Asian values' on the grounds that it is too vague and encompasses too many diverse societies. Instead, Lee prefers to expound on the 'Confucian values' of Confucian-influenced East Asian societies such as China, Taiwan, Korea and Singapore.

7. Creighton (1997, p.212) explains that 'Although the word *gaijin* can be applied to any non-Japanese person it is most commonly only used for white foreigners, who are conceptualised as 'pure *gaijin*', or 'true *gaijin*'. Other non-Japanese groups are designated differently, the collective noun *gaikokujin* (persons from an outside country) being applied to other Asians and black people.'

CHAPTER 4

Escaping whiteness? Primitivism and the search for human authenticity

Introduction

To be outside whiteness is to be outside the cold and instrumental realm of modernity. It seems an outlandish claim. Yet there are many who appear convinced by it: whites who have ventured into non-white territory and found a better, 'more alive', type of person, a more 'real' way of being, and non-whites who have found in their skin colour a symbol of human authenticity.

If we agree that whiteness is associated with the dominant social and cultural forces of the modern era, then it is logical to proceed to speculate that resistance to these forces may have been articulated in terms of a rejection of, or escape from, whiteness: in other words, that critiques of the mechanical nature of 'Western reason' and of the alienated nature of industrial society are likely to contain an implicit or explicit moment of racialised politics. Central to this moment is the forging of non-whiteness as an identity that is *not* alienated and *not* dominated by instrumental logic. Thus non-whiteness is constructed as an oppositional identity, a site from which to critique modernity. This chapter will explore this contention through a particular form of anti-modern critique, namely primitivism. I will be drawing on some familiar and some less familiar names, evoking the primitivist thought of renowned intellectuals and artists but also drawing on the desires to escape whiteness evidenced amongst 'ordinary people'.

Primitivism is often associated with the kind of individuals and the kind of activities discussed in the first section of this chapter, namely artists and art practice. Indeed, those Western artists who have introduced apparently non-Western forms or subject matter into their work have tended to dominate debate on the subject. This singular focus has produced a small but important body of scholarship on primitivist art (Hiller, 1991; Rubin, 1984; Foster, 1985; Rhodes, 1994). But it has also had a less desirable consequence: *primitivism has been connoted as a specialised activity.* An elite of 'creative geniuses' has been constructed as rebels and resistors against 'Western tradition'. From Gauguin to Picasso to more recent representatives, the primitivist has been offered as a singular – albeit somewhat misguided – individual who is prepared to shock and disturb. It is my contention that this conceit has obscured the ubiquity, the sheer everydayness of primitivism. To suggest that primitivism forms part of everyday life in the modern world may immediately be read by some as an attempt to make what is now the highly familiar structuralist point that the excluded term (in this case the non-white or the non-modern) *necessarily* structures and enables any expression of the dominant term (white modernity). However, I would like to resist my argument being collapsed into such philosophical niceties. My approach is sociological. It is the particular forms of primitivism and the different social experiences and roles of the groups that produce them that interest me, not linguistics. Instead of locating primitivism merely as an activity of the radical intelligentsia, such a perspective helps us to find it at work in different ways across the entire landscape of modernity. Thus the signs of primitivism come to be seen spiralling away in all directions. It is there in popular tales and novels of the Wild West and 'darkest Africa', there in Boy Scout countryside retreats and ceremonies, there in the rise of environmentalism and the attempt to preserve wilderness areas, there in popular music, there in the local food store's selection of organic, 'authentic' produce. If this seems a kaleidoscopic list then perhaps I have succeeded in making my point. The flight from whiteness is part of everybody's experience.

The existence of quotidian forms of primitivism raises questions about just what kind of 'escape' it offers. In their analysis of the routinisation of fantasies of transgression in everyday life, *Escape Attempts: The Theory and Practice of Resistance to Everyday Life*, Cohen and Taylor (1992, p.154) suggest that 'we are increasingly surrounded by an ideology which tries to transform every activity into a potential escape. The mundane world is saturated with escape messages.' Cohen and Taylor go on to portray such 'messages' as enabling people to avoid confronting the socioeconomic limitations of their lives. Thus, these forms of 'resistance' are, in fact, forms of subservience; they contribute to the reproduction of the 'mundane' realities from which escapees are attempting to flee. Such an account would imply that 'everyday primitivism' is the stuff of delusion, a symptom less of adventurous subversion than the empty and gestural politics of the truly ideologically acculturated and/or hopeless. However, the relationship between whiteness and modernity suggests that primitivism is more than vacuous day-dreaming. It is necessarily both in and against the social conditions that enable it, an indication of the absences and failures of modernity. Primitivism may be said to be both modernity's conscience and its mortal enemy. It can act – and has acted – as a far-reaching critique of alienation, of industrialised existence, of instrumental reason, a critique sustained by and within everyday life.

There are four parts to this chapter. Each is a separate case study of primitivist thought, a distinct and particular window on the desire to flee whiteness. In the first section, I will be drawing on some of the more familiar art historical sources of primitivism. In the second, part I turn towards the popular context and appeal of primitivism in Britain. This task is undertaken through the examination of the works and life of someone who appeared to offer an authentic voice of the Canadian wilderness, Grey Owl. Grey Owl was a famous 'Red Indian' in the 1930s. However, after he died it emerged he was nothing of the kind. Indeed, he was an Englishman from Hastings, Sussex. The third section engages the gendered nature of primitivism. Again, a specific example is used to explore this theme. The illustration chosen is the so-called mythopoetic men's movement, a diverse grouping of mostly North American white men who have adopted non-Western and pre-modern mythology in order to overcome a perceived crisis of masculinity. The fourth and final section of the chapter turns to the role of primitivism within non-white identities. This last topic is so vast, and yet so rarely written about, that I shall be confining myself to a few, admittedly limited and idiosyncratic, notes. Drawing largely on representations of Africanness in Africa, I will be indicating that certain forms of non-white identity can be usefully understood as expressions of primitivism and, relatedly, that primitivism is no longer the sole preserve of the Western imagination.

The avant-garde, modernity's barbarous mirror

> We fight like disorganised 'savages' against an old, established power. The battle seems unequal, but spiritual matters are never decided by numbers, only by the power of ideas. The dreaded weapons of the 'savages' are their new ideas. ('The "savages" of Germany', Marc, 1992, p. 98, first published 1912)

> Absolute originality, the intense and often grotesque expression of power and life in very simple form – that may be why we like these works of native art. (Nolde, 1992, p.102; first published 1934)

The two expressionist artists cited above, Marc and Nolde, sought to attack what they saw as the outmoded and alienating nature of European culture. To these two men, and many other radical artists both before and since, it seemed obvious that the place to look for models and inspiration to oppose the modern West was the non-West and the pre-modern. The more non-Western, the further away and the less modern, the better. Such a quest is now so familiar as to be a cliché of contemporary cultural production. Indeed, it can appear to be as old as modernity itself. The first modern writer to have attacked his own society's artificiality by comparing it with the more natural, more authentic, society of 'savages' was a French aristocrat from the early modern period, Michel de Montaigne (1533–92). In his essay 'On the cannibals', Montaigne relies on a construct of Nature as a universal 'good thing'. The French, he claims are no longer at one with Nature, they are alienated from Nature, but

> 'savages' are only wild in the sense that we call fruits wild when they are produced by Nature in the ordinary course: whereas it is fruit which we have artificially perverted and misled from the common order which we ought to call savage. (1993, p.231)

Montaigne was aware that his affirmations of otherness were essentially reactions to French society and had very little to do with the realities of other cultures. 'I do not speak the minds of others,' he wrote, 'except to speak my own mind better' (cited by Todorov, 1993, p.41). The same quality of self-knowledge has not always been so apparent within later generations of primitivists. It is a diverse group, consisting of conservatives as well as revolutionaries. However, of all the users of primitivist discourse, few have been quite so bold and remorseless as that collection of late nineteenth- and twentieth-century 'anti-traditional' artists and intellectuals commonly called the avant-garde. In the arts, the avant-garde is associated with the shift from representational realism to more reflexive and non-naturalistic forms of expression (such as impressionism, cubism, expressionism and abstraction). They are also linked to the desire to break down the barriers that separate art from everyday life and the attempt to question the specialised status of art. Avant-garde production is invariably termed modern (or postmodern). However, when seen in the context of the wider social and economic processes of modernity, the avant-garde's relationship to the modern appears decidedly ambiguous. It is true that the avant-garde emphasised the need for newness and originality in culture, of leaving behind the stultifying conventions of bourgeois tradition. However, it is also true that the most important social force that enabled modernity was bourgeois capitalism. Thus, by positioning itself against those social processes associated with bourgeois capitalism, such as instrumentalism and alienation, the avant-garde was effectively staging an attack on modernity. As befits its highly individualised, 'traditional' approach to the creative process (with unique individuals producing unique works in studio or workshop environments), the avant-garde tended to be at least as appalled by the prospect of 'modernisation' as it was attracted to its radical connotations. It may be objected here that the avant-garde could avoid this dilemma through its identification with other political projects linked with modernity, namely socialism and communism. However, even if we were to believe that socialism and communism have been as powerful in shaping modern ideologies and practices as capitalism we would, nevertheless, be confronted with the fact that the way socialist and communist states undertook modernisation was precisely through those processes – social alienation, rationalisation, instrumentalisation – that the avant-garde pitted itself against. It may be concluded that only by abandoning its claims on humanist social critique could the avant-garde make its peace with modernity. The fascist-inclined futurist movement, despite its primitivist tendencies, may be taken as an example of precisely such a manoeuvre. However, for this very reason the futurists are something of an exception in the history of the avant-garde. As a rule, for the Dadaists, surrealists, expressionists and their more contemporary successors, 'humanity' (or a concept akin to it) and individual creativity are things deemed to be of intrinsic value, things that need to be asserted against the crushing forces of modernity.

Thus avant-garde 'modernism' may be considered as existing in and against social and economic modernity. Indeed, if one looks at the rhetoric and practice of the avant-garde in the twentieth century one will repeatedly encounter its practitioners positioning themselves as modernity's other, as the non-instrumental, non-rational, spontaneous opposition to modernity's regime of regulation and dehumanisation. Since the latter characteristics were understood as attributes of 'Western civilisation', the avant-garde articulated its oppositional status by turning to the non-West. Hence, the avant-garde adopted, and identified itself with, the non-Western in order to position itself as socially and culturally radical. 'The exotic,' as Clifford (1988, p.127) explains, 'became a primary court of appeal against the rational, the beautiful, the normal of the West.' An early exponent, the painter Paul Gauguin, found it necessary to leave France in 1891 and travel to Tahiti in search of the authenticity and simplicity apparently absent within metropolitan life. 'I only desire to make a simple, very simple act. In order to do this it is necessary for me to immerse myself in virgin nature, see no one but savages, live their life' (cited by Jordan and Weedon, 1995, p.320).

Although the desire to appear as the 'savage' other of bourgeois society has animated the avant-garde throughout the twentieth century, it may prove useful to offer a more specific illustration. I shall now turn to the examples of Dada (more specifically, the Dada of Tristan Tzara) and surrealism. Dada and surrealism were movements of the late 1910s and 1920s, and the late 1920s and 1930s respectively. Described by Tzara (1981, p.81), in his 'Dada Manifesto. 1918', as the 'abolition of the future', the 'absolute and unquestioning faith in every god that is the immediate product of spontaneity,' Dada was routinely positioned by its supporters as a cleansing process, an attempt to return to a purer, more organic, form of cultural conduct. 'We must sweep away and clean,' noted Tzara in the same manifesto, 'affirm the cleanliness of the individual after the state of madness.' Dadaists sought the destruction of bourgeois art and society, a society whose instrumentalism, rationality and alienated nature had, as they saw it, enabled the mass slaughters of the First World War and was everywhere subverting humanity and creativity. Tzara reached back to the early modern period as the start of the process of Western alienation: 'We want to continue the tradition of the Negro, Egyptian, Byzantine and gothic art and destroy in ourselves the atavistic sensitivity bequeathed to us by the detestable era that followed the quattrocento' (1992, p.63; first published 1919). To cite Tzara again, in his 'Note on Negro art' (1992; first published 1917), 'My other brother is naive and good, and laughs. He eats in Africa or along the South Sea Islands ... From blackness, let us extract light. Simple, rich luminous naiveté.' Lapsing into primitivist aphorism, Tzara continues 'Art, in the infancy of time, was prayer, wood and stone were truth. In man I see the moon, plants, blackness, metal, stars, fish' (p.58). One of the ways Tzara's vision found expression was through Dadaist 'Negro soireés'. These events, according to Maurer, (1984, p.536), were a 'riotous mixture of nonsense poems, sound poems, recitations in various languages, piano playing, drums, jazz, African music and poems, African-inspired masks, screams of laughter, pantomimes, and insults.' As Tzara (cited by Maurer, 1984, p.540) recalled, Dada sought to join 'African Negro and Oceanic art with mental life ... by organizing NEGRO

SOIRÉES of improvised dance and music. It was for DADA a matter of recapturing in the depths of consciousness the exalting sources of the function of Poetry.'

An attachment to emotional release, spontaneity and the evocation of wild freedoms was enacted by the Dadaists as a form of escape from Western white identity. In her article 'The Dada event', Harriet Watts (1988) places such practices within a wider milieu of primitivist celebration. More specifically, she cites an interesting example of a dance event staged in the summer of 1917, and later described by Rudolf Laban, whose school of dance in Zurich provided performers for Dadaist soirées. The particular performance depicted by Laban quickly spilled over into the terrain of ritual. Drawing on a mixture of concocted European folklore and tribal imagery, the event appears to have been designed to access the deepest, least civilised, levels of the spectators' consciousness.

> Shortly before midnight … A group of dancers collected the spectators; torches and lanterns lit the way to a mountain top; overhead, bizarre cliffs looked down on a circular meadow. Here five huge bonfires had been lit. Hopping around and through the flames was a troupe of gnomes. Then a group of masked dancers appeared. The masks were enormous, covering the whole body and made out of stick and grass. These various compressed, towering, pointed and jagged forms concealed the witches and demons creeping up from behind, who suddenly attacked and exposed the masked figures and burned their disguises in a wild dance. Around the dying embers of the bonfires there arose a final dance of the shadows. Then the torches were relit and the dancers led this long procession back to its starting point. These spectators, who came from all over the world to see us, had put up with a lot. (cited by Watts, 1988, p.126)

In fact, despite the admission in Laban's last sentence, spectators were obliged to turn up again, at 6 a.m. the following day, for a dance 'to morning'. The integration of natural, rugged landscapes with the evocation of tribal rituals offered Laban's audience a cathartic, transcendental escape from the mundane world of normal life. However, this performance might also prompt us to consider the way the anarchic, spontaneous elements within primitivism often exist side by side with, and can easily be dominated and displaced by, more authoritarian currents. The latter tendencies draw on the disciplinary, strictly choreographed, aspect of ritual performance, as well as on its tendency towards dazzling audiences through spectacle. Indeed, Watts makes the point that a decade and a half after Laban's overnight spectacular, 'the Nazi party rallying in Nuremberg was to find the torch lit procession just as effective' (p.126).

Unfortunately, the ability to identify the political implications, or even the presence, of primitivism within the avant-garde has consistently been undermined by the latter's tendency to privilege the aesthetics of extremism over all other considerations. To appear 'angry' with the West, to strike a posture of 'refusing' the 'white world', has been conflated with anti-racism. It is an equation that other political radicals and critics of white supremacy have often also appeared to accept. Indeed, it is interesting to observe that a recent edition of the radical anti-racist journal *Race Traitor* (1998, the subtitle of the journal is *Treason to Whiteness is Loyalty to Humanity*) is dedicated to the work of the surrealists. The front cover announces 'Surrealism: Revolution Against Whiteness'. The editor of the issue, Franklin Rosemont (1998a, p.27), identifies surrealism as 'a veritable *school of race traitors*'. It is certainly true that the surrealists enjoyed identifying themselves with non-whiteness and positioning themselves as the mortal enemies of the West. Louis Aragon declared in 1925:

> First of all we shall ruin this civilization ... in which you are molded like fossils in shale. Western world, you are condemned to death. We are the defeatists of Europe, so take care – or, laugh, laugh at us. We shall make a pact with all your enemies. (cited by Rosemont, 1998b, p.6)

In the same year, the Surrealist Group offered a similarly ringing declaration:

> [W]e want to proclaim our total detachment from, in a sense our uncontamination by, the ideas at the basis of a still-real European civilization ... Wherever Western civilization is dominant, all human contact has disappeared, except contact from which money can be made ... the stereotyped gestures, acts and lies of Europe have gone through their whole disgusting cycle. (cited by Rosemont, 1998b, p.7)

The surrealist map of the world privileged those regions most closely associated with mystical otherness and exotic peoples. Tanguy's literal rendering of this map, in 1929, represented Europe (except Celtic Ireland and Russia) as tiny, whilst Oceania and Alaska and Labrador in North America (associated with Native Americans by the surrealists) are hugely inflated (Figure 4.1; see also Roediger, 1998b). The rather famished proportions of Africa in Tanguy's map indicate that, by 1929, the recent fashion for things African had diminished the charm of the continent for the surrealists. Indeed, the foreign objects and areas subjected to the surrealists' gaze were both adaptable and replaceable. Like many other primitivists, the surrealists were not ultimately as interested in non-European peoples as their rhetoric might imply. The latter were employed as symbols, as objects constructed and deployed within a Western debate. In unreflexively fusing libertarian romanticism and racial science to designate 'other peoples' as 'natural', 'timeless', 'static', 'profound' and 'unsophisticated', primitivists invariably locate themselves, albeit unselfconsciously, in a colonial relation to their object of admiration. The latter may be cast as 'more real than the West', as 'more natural than the West', but it is the primitivist, and not the savage, who has the knowledge and the sophistication to designate and describe these characteristics. It is the primitivist who classifies, who discovers, who accords 'respect'. Thus, although primitive people were seen as 'inspirational' by the Dadaists and surrealists, it was an inspiration that could be translated into radical culture only by the creative geniuses of the avant-garde.

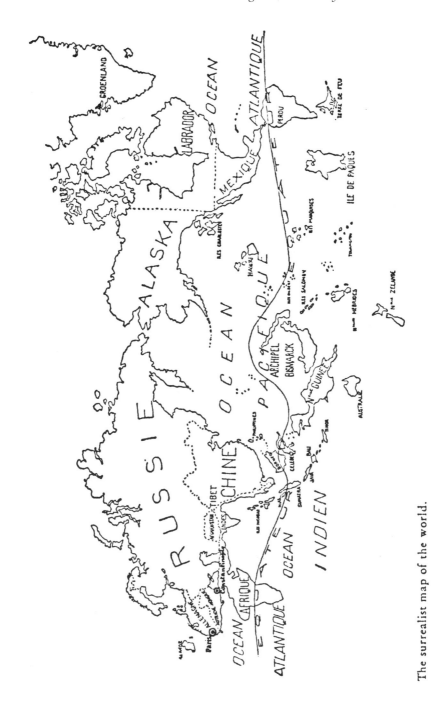

Figure 4.1 'The surrealist map of the world' (1929), attributed to Yves Tanguy

There is nothing quite so dated as an old avant-garde, and few contemporary artists or creative workers would not squirm at the kind of explicit primitivism found within Dada or surrealism. Indeed, over the past few decades, some art critics have claimed that 'our postmodern era' has seen a collapsing of the oppositions Western/non-Western and civilised/primitive. Writing in 1985, Hal Foster (1985, p.206) asserted that 'there are now few zones of "savage thought" to oppose to the western *ratio*, few primitive others not threatened by incorporation.' The 'historical incorporation of the outside' (p.203) is identified by Foster with 'the eclipse of otherness'. Marianna Torgovnick (1990, p.37) makes a similar point on the same theme:

> In the deflationary era of postmodernism, the primitive often frankly loses any particular identity and its sense of being 'out there'; it merges into a generalised, marketable thing – a grab-bag primitive in which urban and rural, modern and traditional Africa and South America and Asia and the Middle East merge into a common locale called the third world which exports garments and accessories, music, ideologies, and styles for Western, and especially urban Western, consumption.

Torgovnick supports this view by reference to a specific incident of art works being made and marketed with an eye to the primitivist market:

> The 'modern'/'primitive' circuit runs both ways now; it's complete. The primitive has in some ways always been a wilful invention by the West, but the West was once much more convinced of the illusion of Otherness it created. Now everything is mixed up, and the Other controls some of the elements in the mix ... in this trade, African carvers gleefully imitate the production of neighbouring traditions, traditions not their own, if those are kinds of statues or masks sell to Westerners; reports have it that the village workshops consult the 'latest coffee-table books and auction catalogues of African art' in order to keep up with shifting Western tastes and vogues. 'Junk' pieces successfully passed off as 'danced' in tribal rituals or, even better, as 'old' will produce many times their usual worth. (Torgovnick, 1990, p.38, with citations from Lemann, 1987)

Foster and Torgovnick both point out that primitivism was once produced in the West and for the West. They are suggesting that today something has changed in this dynamic; that the reference points for who is and who is not primitive have melted away, with the primitive object becoming integrated into the global, consumer-led logic of postmodernity. The question of non-Western primitivism is, I would suggest, the most convincing and interesting aspect of this argument, and I shall be coming back to it at the end of this chapter. However, other elements of Foster and Torgovnick's thesis deserve a somewhat more suspicious treatment. Like many others convinced by the novelty of 'our postmodern era', Foster and Torgovnick overlook the fact that the processes they depict were also apparent within previous periods. The commodification they mention commenced centuries ago. Not unrelatedly, the existence of great chunks of humanity untouched by the West, of people who are not in some kind of relationship with the West, has been doubtful for many generations. It is also pertinent to note that most primitivists have long understood the West to possess its own primitive roots and cultures (note the size of Ireland on the surrealist map of the world), whilst they themselves have often been classed, by others and themselves, as 'savages' within the West. As this all implies, primitivism has *traditionally* attempted to confuse the boundaries between the West and the rest. Yet, as the history of primitivism shows, such transgressions should not be confused with any necessary challenge to colonial or racist stereotypes. However, perhaps the most troublesome aspect of Foster and Torgovnick's position is that they appear to be arguing that primitivism has been eroded as an ideology because Westerners no longer *believe* that there are races less advanced, more primitive than themselves. 'Now everything is mixed up' (Torgovnick, 1990, p.38): Eurocentrism and white supremacism have vanished, it seems, into the circuits of global capitalism. There is an extraordinary blindness at work here to the maintenance of Eurocentrism and the interconnections between postmodern globalisation and racialised discourses of Westernisation (see Chapter 3).

Nevertheless, it does seem clear that the nature of primitivist discourse changed in the second half of the twentieth century. For one thing, a much wider swathe of the population found itself able to 'buy into' primitivism. For, although an interest in primitive peoples has long been widespread, the ability to acquire 'ethnic products' (whether in the form of clothes, food, art and so on) and travel as a tourist to exotic destinations were once almost entirely elite pleasures. Now, by contrast, they are widely available. Perhaps not unrelatedly, amongst today's cultural avant-garde overtly ethnic goods are increasingly regarded as clichéd, even frowned upon as kitsch, and primitivism itself cast as passé and held up to ridicule. And yet, although embarrassed by its colonial connotations, the contemporary avant-garde is finding it difficult to banish primitivism. Primitivism remains, after all, part of the logic of the modern era. It cannot be overcome, or wished away, without a simultaneous overturning of the relationship between modernity, social alienation and instrumental reason. As long as avant-garde art continues to root its transgressions in claims to spontaneity, irrationalism and 'wildness', it will find itself dogged by primitivism.

It might be objected here that many contemporary avant-garde artists employ new technologies as part of their work and, hence, are closer to technophilia than primitivism. That there is no necessary contradiction between the two is indicated by the fact that the Dadaists and surrealists had the same fascination. Moreover, again like the Dadaists and surrealists, even the most virtual of postmodern artists tends to offer a mutant, self-consciously bizarre rendering of the possibilities of technology. A central characteristic of this process is what may be termed 'techno-primitivism'. An example may help to clarify what I have in mind.[1] The Australian performance artist Stelarc has become famous for his interest in technology. I first encountered Stelarc in my home city of Newcastle. Walking up Westgate Road I caught a glimpse of him inside an art gallery (Stelarc and Linz, 1996). He was naked, although a multitude of surgical-looking wires dangled from his arms, torso and left leg. A metallic visor partially covered his face. Once inside, visitors were invited to touch animated representations of the artist on computer screens. These fed electric impulses into his muscles, causing endless spasms of eccentric movement in his wired limbs. Sound sensors on his body were amplified around the room, creating a grinding mass of industrial noise. Other Stelarc projects have involved him having his skin pierced with flesh hooks attached to wires and being winched out from a window to dangle over a street in New York. A more recent work has seen him implant a false right ear under his skin, just below his real ear, a device that can respond to those who whisper into it. It seems clear that the body-piercing and modifying elements of Stelarc's work draw on the symbolism of tribal initiation rites, more specifically evoking the attempt to erase consciousness of the body through mortification. Stelarc's self-consciously grotesque attempts to intertwine cable with veins, electricity with blood, are striking precisely because they appear to be providing a techno-primitivist critique on the bland, safe and bloodless nature of modern life. It is a process that seems to offer an escape from the normal, European body into more extreme, more exotic, realms of experience.

The role of the artist as social outsider has been sustained throughout the twentieth century. It is an orthodoxy that has maintained an image of cultural workers as critics of society, as radicals. As I have sought to indicate in this section, it is also a convention that has traditionally relied on an ability to deploy primitivist discourse as a critique of modern society. For some, the whole idea of 'the primitive' is now redundant, a feature of early twentieth-century culture, an anachronism within 'our postmodern era'. I remain to be convinced. Primitivism is not an epiphenomenon of modernity. It is part of its logic. In the remaining sections of this chapter I will have cause to return to this theme.

'There never came a redder Red Indian to Britain': Grey Owl and the ordinariness of primitivism

[He was] the first Indian [I had met] that really looked like an Indian – an Indian from those thrilling Wild West days of covered wagons, buffalos, and Sitting Bulls. The stamp of his fierce Apache ancestors showed in his tall, gaunt physique, his angular features, his keen eyes, even in his two braids dangling down his fringed buckskin shirt. (Lloyd Roberts's description of meeting Grey Owl, cited by Dickson, 1976, p.214)

A crowd of more than two thousand people heard Grey Owl speak at Sheffield ... no seats remained. In this great industrial city of half-a-million people his talk of clean air and water had a great impact. This city of factory chimneys smelled of sulphur. The shallow river that ran through the town was usually bright yellow, awash with chemicals ... Grey Owl repeated his central theme: 'The difference between civilised

man and the savage is just this – civilised people try to impose themselves on their surroundings, to dominate everything. The Indian's
part of the background. He lets himself – not just drift – but go with
Nature.' (Smith, 1990, p.120, with citation from Grey Owl, 1936)

The desire to escape the confines of modern society is not restricted to a
tiny elite of artists. Indeed, it may be argued that since it is ordinary
people who bear the burden of an alienated, nine-to-five, industrialised
existence, it is they who are most likely to search for release in the pleasures of temporary escape, whether that entails a holiday to an 'exotic'
destination or the ability to 'get completely out of it' in the countryside, in
a film or video game or, indeed, in a bar. The desire to consume the primitive, to thrill to tales of perilous excitement in wild landscapes, may be
traced back as far as the advent of commercial 'popular culture' itself. In
Britain, some of the first examples of the mass consumption of primitivist
imagery occur with the reception of books relating colonial adventure.
The mass appeal of non-fiction travel accounts, such as Livingstone's
Missionary Travels (1857), and myriad fictional tales of adventure is testament to the widespread sense that non-Western destinations offered the
kind of fun, the kind of beauty, as well as the kind of horror, not available
in Britain. Such accounts are sometimes claimed to have inspired other
travellers and instilled in readers a wish to further Britain's imperial
glory overseas. But there is another side to such eagerness for departure,
the story of why many ordinary people wanted to get out of Britain
despite an uncertain future elsewhere. For some, no doubt, it was a
wrench to leave, and there is plenty of evidence that many attempted to
shape their colonial homes as far as possible in the image of the 'home
country'. Yet for others, the colonies were a source of escape, an escape
from a society they found stultifying and confining. It is probable that
stories of Britons who 'went native', abandoned the idea of 'return' and
completely crossed over into the other's territory, are more important for
their ubiquity, and the way they tantalised and excited the imagination of
imperialists and settlers, than for their veracity. However, many Britons
do appear to have taken at least a few steps along this path. For most, no
doubt, the attraction of the cultures and peoples of non-European lands
was not stimulated by any conscious desire to abandon being white but
rather from a pressing sense of the limitations of life in Britain. This was
initially the case with the man who I shall be using to explore the popularity of primitivism, Grey Owl.[2]

Grey Owl was a hoax. Or at least that is what the English press proclaimed him to be in 1938, the year of his death and the year that his 'real
identity' was revealed. '"GREY OWL" WAS NOT A RED INDIAN – HE
WAS A SUSSEX MAN' proclaimed the headline in the local paper of the
town where he was born, the Hastings *Argus* (cited by Smith, 1990, p.211).
The 'half-breed' noble savage who had undertaken two successful tours
of Britain, giving talks about 'Indian' life to large crowds, was exposed as
Archie Belaney, born in 1888 in England. His native attire, his pony-tails,
his mysterious and proud 'Indian features', were 'fakes'. People had been
fooled, and they had so much wanted to believe.

Figure 4.2 Grey Owl feeding a beaver kitten, Prince Albert National Park, from L. Dickson, *Wilderness Man: The Strange Story of Grey Owl* (1976)

By the time of his death, Grey Owl was widely seen as a principal exponent of and for 'the wilderness' (Figure 4.2). His four popular books of native anecdote and lore offered clear conservationist messages that directly challenged the wisdom of sacrificing Nature in the service of development. Today *The Men of the Last Frontier* (1931), *Pilgrims of the Wild*, (1935), *The Adventures of Sajo and her People* (1935) and *Tales of an Empty Cabin* (1936) are regarded as classic statements of 'green philosophy'. Grey Owl's works were reprinted in the early 1970s. Moreover, his cabin, Beaver Lodge in Saskatchewan, was restored in 1988, to be made the centrepiece of a wilderness sanctuary. Francis (1992, p.140) has argued that the 'greening of the Indian begins with Grey Owl. To him belongs the credit for affirming, if not creating, the image of the Indian as the original environmentalist, an image which has gained strength in the years since he expressed it.' Nevertheless, the public acclaim of today cannot bear comparison with the excitement and eagerness with which Grey Owl was received in the 1930s. As described by his publisher Lovat Dickson (1976):

His appearance in London in October 1935 created a sensation. Not only did he look romantic, he spoke pure romance. His thrilling voice brought the wilderness and its inhabitants, animal and Indian, alive for audiences. He won the hearts of everyone from the monarch to the smallest child who heard his stories and saw the films which the Canadian government had made of his cabin in the woods and of the animals who lived in and around it with him. (pp.4–5)

The 'fakery' of Grey Owl might be regarded as mendacious. But his story is more revealing, and better understood, when it is realised that Grey Owl was not a natural fraudster. A boyhood friend from Hastings, Margaret McCormick, mused many years after his death that she still found it difficult to know whether Archie Belaney became Grey Owl or Grey Owl became Archie Belaney (Dickson, 1976). From his earliest days in Hastings he had felt alienated from England. A sense of dissatisfaction with the country was something of a family trait. Grey Owl's wayward and alcoholic father, George Belaney, despised the monotony of English life. His real passion was Nature, more particularly for wild animals, which he hunted, keeping and stuffing his kills for display. Grey Owl was four when his father was offered a stipend by other family members on the condition that he left England and never returned. George Belaney went to live in Mexico, where he died twelve years later. A sense of the smallness and emptiness of life in England also appears to have been present in Grey Owl's early life. He spent many hours tracking animals, alone in the woods around Hastings, playing at being an Indian. The   official history of his school, the grammar school, records that

> he was a delicate boy but full of devilment; and fascinated by woods and wild animals … What with his camping out, his tracking of all and sundry, and wild hooting, he was more like a Red Indian than a respectable Grammar School boy. (cited by Dickson, 1976, p.28)

Grey Owl's veneration of nature and his identification with 'Red Indians' were not learned in Canada but in Hastings. The horror of industrialisation and love of the countryside are, of course, familiar and, supposedly, typical characteristics of the English. Archie was not considered a strange child for wishing to commune with Nature or because he liked playing at being an Indian. Nor were his activities entirely solitary. Another school friend recalls that 'he ran the Belaney gang, who played at Red Indians all the time' (cited by Smith, 1990, p.16). If we find a echo of his father in Grey Owl's desire to flee England, it his Aunt Ada who seems to have first inspired his love of nature. The dedication in *The Men of the Last Frontier* (1931) reads 'as a tribute to an aunt whom I must thank for such education that enables me to interpret into words the spirit of the forest, beautiful for all its underlying wildness.' Readers would have supposed that the aunt in question was a 'Red Indian', a women of the forests, a 'native of the wilderness'. They would probably not apply any of these terms to the person Grey Owl actually had in mind, namely Aunt Ada, a seemingly conventional middle-class English woman. But, as one learns more about Grey Owl's background, it becomes increasingly apparent that his dislike of the confines of modern life and his wish to search out a pristine rural idyll were not out of keeping with his familial and wider boyhood experience. His attitudes developed within a widespread and everyday culture of popular primitivism. His father, his Aunt Ada, his school friends, did not go on to identify with Native Americans as Grey Owl did. But in their different ways they shared in his project. They too had a sense that Nature needed to be sought out, that excitement could be, and perhaps needed to be, searched for away from the confines of modern life, away from white civilisation.

It is reasonable to surmise that Grey Owl's primitivism was also influenced by the presence of themes of wilderness, of primitive people and exotic escape, within more public aspects of British popular culture. The identification of the Red Indian with wilderness had found an influential voice early in the ninteenth century with James Fenimore Cooper's series of novels, most famously *The Last of the Mohicans* (1826). Cooper's hero, Natty Bumppo, was in 'flight from civilized unmanliness to native-American traditions of patriarchal comradeship' (Leverenz, cited by Kimmel, 1995, p.123). His works were a particular favourite of the young Archie Belaney, as was Longfellow's poem *Hiawatha*. Another American author Archie enjoyed was Ernest Seton. Seton wrote twenty or so books on the theme of animals and the love of Nature, his first work being *Wild Animals I Have Known,* published in 1898. Archie Belaney seems to have been especially keen on another best-seller by Seton, *Two Little Savages: Being the Adventures of Two Boys who Lived as Indians and What they Learned* (1903). The novel is full of practical tips explaining how one may build a teepee, read smoke signals, light fire with sticks and so on.[3] Donald Smith (1990, p.18) suggests that 'Archie might well have patterned his future life on *Two Little Savages.*'

The late nineteenth century and early twentieth century saw the publication of numerous testimonies and tales from the 'American Wild West' in Britain, as well as variously sympathetic accounts of 'Red Indian' life. What lent the latter their particular appeal was, in part, their pathos. The Red Indian was considered part of a 'dying breed', a noble remnant of a culture that was bound to 'disappear' through contact with the superior and stronger civilisation of white settlers (Gidley, 1992). Thus politically neutered Red Indians could become objects of sympathy, although their savage nature was rarely placed in doubt. Moreover, popular interest was consistently framed by a kind of excited horror that such strange people could exist at all. The visit by Buffalo Bill and his Wild West Show to Hastings in August 1903 offered audiences – among whom was the young Grey Owl – a variety of thrills, not least of which was the spectacle of seeing tomahawk-wielding, bloodthirsty Red Indians in the flesh. Subsequent to this visit a craze for Buffalo Bill swept through Belaney's school, as it had through many others. A few years later, after Belaney had left England, a similar enthusiasm was drawn upon in the Boy Scout movement. The movement's leader, Baden-Powell, organised weekend camping expeditions for boys. During these retreats into the 'wilds' boys wore Native American headdress, and emulated native tracking skills and other ceremonies. The Boy Scouts employed a variety of different native cultures in order to organise their displays of masculine prowess. References to India and Africa, as well as Native America, were all deployed in order to symbolise the acquisition of knowledge of Nature, both animal and human. This tendency to mix together and homogenise 'native' cultures meant that the popularity of things Red Indian in Britain often overlapped with what were understood as similar cultures, similar peoples in other parts of the world. Thus, whether it was Tarzan 'revelling in the freedom of the fierce, wild life he loved' (Burroughs, 1963, p.149), or Kim's ability to integrate into any aspect of the culture of India (more particularly to dress like and, hence, become a native boy: 'a demon woke up and sang with joy as he put on the changing dress and changed speech and gesture therewith'; Kipling, 1912, p.226), it was not the cultural specificities of the other that mattered so much as the act of racial transgression itself. More precisely, it was the possibility that whites could and did escape from whiteness, that they could choose to adopt (and by implication, at any time choose to discard), the ways of 'the tribal', 'the natural', the 'less sophisticated' peoples of the Earth.

Not long after he first settled in Canada (in Temagami in northern Ontario), at the age of 17 in 1906, Grey Owl appears to have begun to change his appearance. He grew his hair long, he wore native clothes, he developed a native bearing, the 'stern and wise' facial expression that Donald Smith (1990) calls his 'Indian face'. He also taught himself native languages, lore and trapping skills. When precisely this cultural identification spilled over into the fabrication of a new family history is not entirely clear. However, his British publisher Lovat Dickson comments that Grey Owl's feelings of empathy for 'Indians' were considerably strengthened as he came to consider his new home, in Temagami, as being under threat from white civilisation. As Dickson explains,

> Archie's romantic dream of living with the Indians in isolation was shaken by the rumours of this paradise becoming a white man's holiday ground, and the Indians becoming servants in surroundings in which they were nature's master. It made him despise what he had come from, and romanticise the more what he had come to. (1976, p.71)

Belaney's 'Indian' identity was designed for a white audience. He felt that his written work would be better received, and that his pleas for the preservation of Canada's wilderness landscapes and people would be better heeded, if he were taken to be a native. As this implies, the real moment of transition from Archie Belaney to Grey Owl was prompted by the knowledge that people wanted to hear from a 'real Indian', that there was a pre-existing disposition towards finding such people remarkable. His second published essay (which was, like the first, sent to *Country Life*) was accompanied by a letter (cited by Dickson, 1976, p.181) that adopted a self-consciously uneducated style. 'I guess I shouldn't be talking so personally to you, Mr. Editor, being a business man,' he wrote, explaining that his real name was Wa-Sha-Quon-Asin, or Grey Owl. The literary transformation of Englishman to Indian is detailed by Smith (1990):

> If he wrote as a North American Indian he knew that his public would credit him with an insight into nature denied non-native writers. From Cabano on May 6, 1929, he wrote to Country Life of Indians as 'them', but claimed to have been adopted by the Ojibwa about twenty years earlier. By November 5, 1929, he reported to his English editor that 'for about 15 years he spoke nothing but Indian'. On November 12, 1930, he signed himself 'Grey Owl' for the first time, after once trying out the

name, 'White Owl'. The choice of his new name came easily. Had he not imitated the hoot of an owl since his boyhood days in Hastings? Like him many owls were active at night. On February 2, 1931, the new Grey Owl announced to Country Life that he had 'Indian blood'. Five months later he completed his transformation. On July 1, 1931, he declared himself as an 'Indian writer' who 'writes as an Indian'. (p.85)

Driven, in part, by a desire for fame, Grey Owl increasingly immersed himself in a native identity. Once the details of his new 'Indian' identity were fixed Grey Owl stuck to them, inhabiting his role in both public and private. He was, he said, a 'half-breed', born in Mexico of a Scottish father and a Jacarilla mother. His parents had, he claimed, only ever been to Britain as part of Buffalo Bill's Wild West Show. Grey Owl found himself not merely accepted in this role but considered a 'fine specimen' of the race. Indeed, he seems to have revelled in the part, to the point of what sometimes seems like mockery. To one Glasgow journalist eager to know more about the habits of 'Indians' he explained, 'I love classical music. Wherever you find an Indian camp, you will find a gramophone and a pile of records, most of them Beethoven' (*Evening Citizen*, 28 November 1935). However, the clamour associated with Grey Owl's two speaking tours of Britain (in 1935–36 and 1937) could hardly have been expected. During his first tour he gave 200 lectures in four months, addressing nearly 250,000 people. The second tour, which included a private performance for the British royal family, was equally punishing. 'There Never Came a Redder Red Indian to Britain', enthused the *Sunday Express* (24 April 1938). Dressed in Indian attire, Grey Owl was accepted as the mouthpiece of the wilderness. Grey Owl himself appears to have been quite at ease with this role. 'When I stood on those platforms I did not need to think. I merely spoke of the life and the animals I had known all my days. I was only the mouth, but Nature was speaking (Grey Owl, cited by Dickson, 1976, p.237).

There is a savage irony in the fact that the gruelling, alienating life of being on the road, talking to endless audiences in numerous cities about the uncorrupted simplicities of Nature, appears to have killed Grey Owl. Less than a month after his last lecture, in Toronto, he went into a coma, apparently caused by exhaustion. He died the following day, 13 April 1938.

Grey Owl had become a symbol of sentiments shared by many, feelings of disaffection with 'the modern world', of a desire to escape the confines of industrialised 'white civilisation'. In the aftermath of his death, and the exposés that followed, he was widely condemned in the British press as a fraud. In Canada, though, a more benign attitude seems to have prevailed. Grey Owl had emerged as a new voice for Canada, a voice that provided a sense of national identity based not upon old colonial ties but upon Canada's unique wilderness. Whilst alive he had been fêted as, in the words of the *Canadian Magazine* (1936; cited by Smith, 1990, p.169), 'a great Canadian', a man who challenged 'all good Canadians to remember that their heritage is a heritage of the North and that to forget it is to make ourselves a little people.' The sense that the nation needed a figure like

Grey Owl seems to have coloured the Canadian obituaries written about him. Thus, for example, under a headline 'Fun to be Fooled', the Winnipeg *Tribune* (20 April 1938) explained that

> The chances are that Archie Belaney could not have done nearly such effective work for conservation of wild life under his own name. It is an odd commentary, but true enough, that many people will not listen to simple truths except when uttered by exotic personalities.

However, it was Lovat Dickson who provided the most dogged defence of Grey Owl. In an edited collection of tributes – *The Green Leaf: A Tribute to Grey Owl* (1938) – and in a biography *Half-Breed: The Story of Grey Owl* (1939; the term 'half-breed' was transformed over the generations into *Wilderness Man: The Strange Story of Grey Owl*, 1976, in Dickson's later refashioning of his earlier biography), Dickson tried to show that there was nothing inauthentic about Grey Owl's commitments, either to the environment or to native Canadians. For Dickson it is the public's identification with Grey Owl, their expectation of him, that is the key to his story:

> here was this romantic figure telling them with his deep and thrilling voice that somewhere there was a land where life could begin again, a place which the screams of demented dictators could not reach, where the air was fresh and not stagnant with the fumes of industry, where wild animals and men could co-exist without murderous intent. (1976, p.240)

Landscapes of masculinity: the mythopoetic men's movement

> What is driving men in ever increasing numbers into the woods to beat drums and share grief that they can't articulate in their everyday lives? (Harding, 1992)

The 1980s and 1990s witnessed the development of a 'new men's movement'. Across North America, so-called 'mythopoetic' men's groups and journals attempted to get in touch with the 'deep male', proclaiming 'bold visions in the time of masculine renewal' (Harding, 1992). In this section, I want to look at this movement's beliefs and practices. More specifically, I will be focusing on its adherents' representations of racial identity and wilderness. Briefly stated, I shall be arguing that (1) mythopoetic men are creatively reworking colonialist fantasies of non-Western societies and landscapes; and (2) that this process acts to naturalise the movement's adherents' contradictory experiences of power.

First, I will introduce the movement in greater detail, outlining both its central ideas and the socioeconomic background of its supporters. Second, I will describe and criticise existing academic explanations of the movement. In the third part, I will show how certain colonial representations of 'racial' and landscape difference and identity have been adopted and adapted by its followers. My analysis will focus on two examples of mythopoetic practice: retreats and initiation rites. The latter example will also be used to explore the creative, mutative nature of mythopoetic cultural appropriation.

The mythopoetic men's movement: an overview

There exist today three men's movements: the anti-sexist men's movement; the men's rights movement, which campaigns around issues such as divorce and child custody; and the 'mythopoetic' (or 'new') men's movement. Although it is the latter I am concerned with here, there is considerable dialogue between this grouping and men's rights and anti-sexist activists (see, for example, Kammer, 1992; Brod, 1992; also Hearn, 1993). Between the anti-sexist and mythopoetic traditions, the ideology and practices of consciousness-raising therapy provide the most significant area of shared concern. Indeed, a line of descent may be traced from the new men's movement to the 'therapeutic turn' that took place within New Left politics in the early 1970s. At that time, anti-sexist men's activism, especially in the United States, was dominated by the organisational and intellectual structures of 'masculinity therapy'. These structures included workshop-based counselling and 'emotion sharing', charismatic leader figures and psychological jargon (for accounts of the results of this work see Goldberg, 1976; Ellis, 1976; Lyon, 1977). As later commentators have observed (Connell, 1995), 'masculinity therapy' enabled the development of an individualistic, 'guilt-trip'-oriented, political agenda. However, it also provided a conducive environment for formerly pro-feminist men to develop alternative psychological approaches to gender issues, approaches that, whilst maintaining a focus on the individual, angrily rejected the 'politics of guilt' and asserted the need for men to stop being 'ashamed' of their masculinity. Male mythopoeticism was facilitated by, and may be seen as a development of, this revisionist perspective (Fee, 1992; Connell, 1995). The influence of therapeutic models is also immediately recognisable in the organisation of the movement. Workshops, counselling, gurus and 'psycho-babble' are the stock-in-trade of the 'new primitives'. These structures, which once served anti-sexism, are now employed to celebrate and affirm gender essentialism.

The mythopoetic men's movement first emerged in the early 1980s in North America. From minor beginnings, a significant network of men's groups and related journals gradually grew up. Harding (1992) listed thirteen North American mythopoetic journals (including *Thunder Stick* (Figure 4.3), *Wingspan* and *The Talking Stick*). More recent additions to mythopoetic publishing include *RavenFlight* and the English magazine *Mandrake: Journal of Men, Earth and Spirit*. However, many men have been attracted towards the movement through one book, Robert Bly's *Iron John* (1990). *Iron John* is a heady mixture of reworked folklore, social commentary and poetry. It offers readers a series of social and psychoanalytical insights into the importance of men's ability to free themselves from dependence upon mother figures and embrace their own autonomous masculinity. The 'percentage of adult sons still living at home,' Bly warns, 'has increased.' Moreover,

Figure 4.3 *Thunder Stick* (1992) (front cover)

we can see much other evidence of the difficulty the male feels in breaking with his mother: the guilt often felt toward the mother; the constant attempt, usually unconscious, to be a nice boy; lack of male friends; absorption in boyish flirtation with women; attempts to carry women's pain, and be their comforters; efforts to change a wife into a mother; abandonment of discipline for 'softness' and 'gentleness'; a general sense of confusion about maleness. (Bly, 1990, p.43)

Robert Bly's criticism of the mother-dependent 'soft' male, and his quest for the 'deep' or 'hairy' male, characterises the movement as a whole. His identification of contemporary Western society as feminised also finds many echoes in the primitivist practices its adherents have developed.

For the most part, these practices draw from non-Western (especially Native American) and pre-modern cultural forms. Thus cultural forms and values associated with tribal cultures and European folk traditions are utilised to express, or tap into, the unfeminised essential male. The most important of these activities, some of which I shall discuss in greater detail later, include group drumming, group incantation, wilderness retreats, men's councils, initiation rites, the (re)creation of male rituals and the invocation of age-based male hierarchies (for example, the use of the title 'elder').

It is important to note, however, that in some parts of the contemporary scene, echoes of anti-sexism may also be heard. Mythopoetic workshops in Britain have included consciousness-raising exercises on pornography and male violence (Crampton, 1994). There also exists an active debate in certain quarters of the movement about the relationship between mythopoetic and political praxis (Brod, 1992; Stothers, 1992; Wolf-Light, 1994). As we shall see later, pro-feminist concerns have also fed into some mythopoetic ritual and retreat work. Indeed, a number of activists have sought to explore what they perceive to be the mutually supportive link between their own praxis and the work of eco-feminists such as Starhawk (1982). Thus, for example, the editor of the Washington State-based magazine *RavenFlight: A Journal of Men's Arts and Mysteries*, describes it as designed 'for spiritual feminist men' (Carrell, 1994a, p.3). Other contributors to the journal have explored the relationship between 'men and the goddess' (Lertzman, 1994) and spoken of 'the Great Mother as underlying everything on the planet ... as being the underlying force that moves the universe' (Chamberlin, 1994, p.36).

Nevertheless, if one looks at more typical mythopoetic material, the predominant tendency is explicitly, or implicitly, hostile to women. Even more characteristic is a disengagement from overt politics, a process that appears to be interwoven with a concern to explore and mythologise the 'deep structures' of the male psyche.

The following account, of the final moments of the Clearwater Men's Group's weekend retreat (held in Wisconsin in September 1991) provides a glimpse into the movement's favoured mode of physical and verbal expression.

> We would celebrate a ritual of our own making, a ritual of striving, ascending, and joyous welcoming. We broke from the circle to plant our staffs in a double row, forming an aisle running fifty feet up a steep hillside. At the top of the hill we began to drum, building quickly to a throbbing boogie. Then one by one, beginning with the eldest, we descended to the bottom of the hill, entered the path between the staffs and walked, ran or danced our way to the top to be hugged, held, and hoisted into the air by our brothers' strong arms. As the youngest man reached the hilltop, the drums built up to a new climax and we raised a triumphant chant: 'WE ARE MEN!' (Pierson, 1992, p.114)

The similarities of this event to the Dada-influenced dance ritual staged by Laban, cited earlier, are striking. Both include a ritualistic climbing of a hill, role-play and chanting, and cathartic incantations of solidarity.

Moreover, in both cases one finds an overt display of emotion and pleasure, a reworking of apparently non-Western rituals, symbols and structures of authority in a rural, 'wilderness', context. These themes are woven together into a poetic and mythic performance, a performance that is enacted through a stylised repertoire of language and action designed to connote a primitive, pre-industrial, way of being. Of course, for the mythopoetic group another element in this celebration is also central, a celebration of masculinity.

So what kind of men belong to this new formation? Several commentators within the movement have noted, often with disappointment, that mythopoetic men are overwhelmingly white and middle class (Carrell, 1994b; Daly, 1992a). The sociologists Kimmel and Kaufman's (1994) more detached observations of what they refer to as 'a variety' of mythopoetic events support this conclusion:

> Attendance of men of color ranged … from zero to less than 2%, while never greater than 5% of the attendees were homosexual men. The majority of the men were between 40 and 55, with about 10% over 60 and about 5% younger than 30. Professional, white-collar, and managerial levels were present in far greater proportion than blue-collar and working-class men. (p.263)

These assertions, although by no means conclusive, clearly beg further questions. Why are men, and white middle-class men in particular, apparently being attracted to the pleasures of gender essentialism? And why now? In the next section, I offer a critique of the existing responses to these two questions.

Explaining the mythopoetic movement

The media's treatment may be characterised as a discourse of derision. It has consistently portrayed mythopoetic events as ridiculous and bizarre: 'a depthless happening in the goofy circus of America' (*Time* magazine, quoted by anon., 1992). The non-journalistic assessments that currently exist are less overtly mocking but often equally negative (Hagan, 1992; Kimmel, 1995; Schwalbe, 1995). Kay Hagan's (1992) edited collection *Women Respond to the Men's Movement* presents a series of attacks on the patriarchal and hierarchical agenda of the 'new primitives'. 'This "men's movement",' Margo Adair (1992, p.55) notes in her contribution, 'is not about social change. It is a backlash – men clamouring to reestablish the moral authority of the patriarchs.'

The notion that mythopoeticism represents an attempt to reaffirm and reinstate male power, and undermine the gains that have been made by feminism, finds an echo in sociological appraisals by Richard Collier (1994), E. Anthony Rotundo (1993), and Michael Kimmel and Michael Kaufman (1994; see also Kimmel, 1995). However, these authors also propose more complex and historically nuanced explanations. They are especially concerned to emphasise (1) the movement's relationship to a contemporary crisis of masculinity (a point emphasised by Kimmel and Kaufman); and (2), the resonances of context and content between the movement and masculinist projects of the late nineteenth and early twentieth centuries (this reading is emphasised by Collier and Rotundo). I will outline each of these analyses in turn, beginning with the former.

Kimmel and Kaufman argue that late twentieth-century socioeconomic restructuring has created the preconditions for the mythopoetic movement. More specifically, they suggest that an erosion of middle-class males' traditional social and economic authority has created a pool of bewildered and 'threatened' males, men in search of the certainties of traditional gender identities. Kimmel and Kaufman focus on two processes: (1) the decline in middle-class males' economic autonomy (for example, 'increased bureaucratization of office work', p.261); and (2) the erosion of the exclusivity of this group's traditional roles (for example, the increased gender, 'racial' and sexual heterogeneity of white-collar occupations). They argue that these changes have hit 'middle class, straight white men from their late 20s through their 40s' the hardest because these were the men with most status invested in traditional power relations. These men have 'experienced workplace transformation as a threat to their manhood and the entry of the formerly excluded "other" as a virtual invasion of their privileged space' (p.262).

Kimmel and Kaufman's emphasis upon the subversion of class and gender roles is a helpful corrective to the relentlessly individualist and self-mythologising histories of the movement developed by its adherents (Harding, 1992). However, their analysis is also marked by a number of unsatisfactory assumptions concerning social causality. In particular, the assertion that white heterosexual middle-class men have 'experienced workplace transformation as a threat to their manhood' more than any other group of men (a contentious point which is not empirically supported by Kimmel and Kaufman) and, *therefore*, have become the core of the new men's movement, represents an unhelpful simplification. It is an analysis that suggests that socioeconomic change has an obvious, commonsense, effect and meaning. Indeed, Kimmel and Kaufman make no mention of any of the sociological literature on the emergence of political consciousness amongst middle-class groups. Most damagingly, the vast body of work on the development of what, to adopt a somewhat dated but useful expression, may be called 'post-materialist' values amongst sectors of the middle class is neglected. Numerous studies of this sort have indicated that public professional, creative and other 'nontraditional' elements of the middle class have at their disposal interpretative structures, educational resources and activist and political traditions that emphasise personal growth, self-expression and environmentalism (see, for example, Eckersley, 1989; Inglehart, 1977; 1981). Since it appears to be these class elements and these concerns that dominate the mythopoetic movement, I would propose that the movement's formation has been facilitated by a process of post-materialist interpretation of socioeconomic change, rather than simply by the fact of economic change itself.

As this suggests, Kimmel and Kaufman's assertion that socioeconomic restructuring 'threatened' white heterosexual middle-class males more than any other group is misplaced. For, more threatened or not, this group contained within it interpretative traditions capable of reading and responding to change in particular ways. As this implies, the absence of a working-class or non-white mythopoetic men's movement says more about the nature of contemporary political activism and cultures than the absence of workplace transformation or a crisis of masculinity amongst these groups.

Not unrelatedly, the portrait provided by Kimmel and Kaufman of the bewildered middle-class male, a figure pathetically scrabbling around for traditional comforts in a turbulent world, fails to convey the dynamic and inventive (and, as media ridicule suggests, highly unorthodox) process of cultural re-inscription at work within this new men's movement.

The extraordinary masculinist assertiveness of the mythopoets clearly reflects fears about male powerlessness. However, it is also a testament to middle-class men's experiences of, and ability to wield, power: of their knowledge that they, unlike so many other groups, can shrug off ridicule; that they can obtain considerable resources for their projects; that they can take unguilty pleasure in the act of *play*. In other words, the mythopoetic movement reflects its adherents' contradictory experiences of power.

Kaufman (1994) has written of this contradictory experience as a fundamental component of modern gender relations. 'Men enjoy social power,' he notes, but 'the way we have set up that world of power causes [them] immense pain, isolation, and alienation' (p.142). However, more specific forms of contradiction may also be isolated. For example, within the workplace we see both affirmations and subversions of male power. On the one hand, the increasing role of women, in previously exclusively male positions, has undermined men's status. And yet, on the other, any sense of loss that may have arisen from this process needs to be set against the fact that the actual amount of power wielded by men, *particularly those from the affluent middle class*, is still enormous. The leading positions within the economy, as well as within many other spheres, are, after all, still occupied by men. Indeed, according to one *Thunder Stick* contributor, although men 'do not feel powerful', and 'carry a malaise', nevertheless, it must be admitted that 'white middle class males are the most powerful social class to ever walk the face of the earth' (Stothers, 1992, p.20).

The men within the mythopoetic movement also appear to experience both power and a sense of a loss of power through their relationship to feminism. Many of the movement's adherents have written and spoken about the way feminist interventions have provoked self-doubt and emotional crisis within their lives (for example, Pierson, 1992; see also Kammer, 1992). However, this experience of challenge, with its concomitant loss of masculine authority, should not be exaggerated. Feminism has unnerved many men, but its power must be placed in the context of competing masculinist agendas and initiatives. Mythopoetic men have no shortage of masculinist archetypes to emulate (for example, in Hollywood films and on television), or mainstream manly activities to engage in (for example, most sports and leisure activities). As this implies, for men, being threatened by feminism in a masculinist society is to encounter conflicting sensations of power.

I will be looking at the way these contradictory dynamics are played out in mythopoetic practice later in this chapter. I want to turn now, though, to the second of the two sociological approaches mentioned earlier. Like Kimmel and Kaufman, Collier (1994) and Rotundo (1993) rely heavily on the notion of a contemporary crisis of masculinity. However, both also emphasise the historical resonances of recent men's movement activity with masculinist projects of the *fin-de-siècle* (see also Kimmel, 1995). Indeed, Rotundo claims 'that the spiritual warrior of today closely resembles his masculine-primitive ancestor of a century ago' (p.288). The 'ancestor' in question was reacting, explains Collier, to the rise of the nineteenth-century women's movement and an emergent 'moral panic' around the notion of male homosexuality (see also Showalter, 1992). In response, new, aggressively affirmative, visions of manhood were developed, visions that, to quote *Century Magazine* (1896, cited by Rotundo, 1993, p.224), called for a 'vigorous, robust, muscular Christianity ... which shows the character and manliness of Christ.' This gender model fed into the creation of the Boy Scout movement and the rise of athletic clubs. It also accompanied and sustained the idealisation of military training and armed combat, practices extolled as essential to the 'new manliness' (Crosby, 1901, p.874).

Rotundo locates the most dynamic adherents of late nineteenth- and early twentieth-century notions of 'new manliness' within the middle class – more specifically, that group of middle-class men who were experiencing a reduction in their status and employment mobility because of the feminisation, bureaucratisation and rationalisation of white-collar occupations. In a formulation that echoes Kimmel and Kaufman's observations on a later generation of disgruntled middle-ranking males, Rotundo suggests that the 'new work world of the middle class threatened manhood' (p.250) and, hence, provoked a masculinist backlash.

It seems that, despite its crudeness, this reductive chain of social causation has become central to much recent writing on masculinity. However, sociological simplification is not the only problem with this explanatory model. It also, erroneously, tends to cast 'men's movements' as entirely negative and reactive creations; truculent and unimaginative assertions of orthodoxy in a changing world. I want to argue that the contemporary mythopoetic movement represents more than a simple rejection of gender equality. It is, as I shall now show, also an attempt to reconstruct a 'traditional' male identity using reworked myths of 'racial' and landscape difference from the colonial era.

Primitive people, primitive landscapes: an analysis of mythopoetic practice

O, I am in
the Wildwood
deep, so deep
(Matthews, 1994)

In this section, I shall explore how the movement expresses its gender essentialism through reworked colonialist notions of 'racial' and landscape identity. I shall also argue that this process naturalises the movement's adherents' contradictory experiences of male power. The two examples of mythopoetic praxis that I will be discussing are the retreat and the initiation of 'boys into men'. However, before introducing these examples, some general and historical observations on the movement's intellectual relationship to issues of 'race', landscape and identity may be useful.

The new men's movement's representations of cultural identity may be traced to forms of representation developed within the colonial era. More specifically, the forms of primitivism it makes use of draw from an established and well-supplied reservoir of imperial fantasies of 'tribal' peoples as metropolitan 'civilisation's' Natural Other. Four aspects of colonial representation are particularly relevant to this argument:

1. The notion that authentic tribal societies are unchanging and timeless.
2. The representation of tribal societies as animalistic, 'at one' with Nature and 'of the wilderness'.
3. The notion that gender inequality in tribal societies represents a more primeval and natural form of gender relations.
4. The sense that white men may experience freedom from social constraint, and a sense of liberation, by venturing into wilderness regions.

(For further discussion on these four themes, see Gidley, 1992; Torgovnick, 1990; Brantlinger, 1986; Short, 1991; Spurr, 1992.)

The development of these representational strategies may be traced within a variety of Western cultural forms and societies. The cult of the primitive in European romantic and avant-garde art and literature is perhaps the best-documented and most-debated example. However, the new men's movement draws on a distinctly North American refashioning of colonist fantasies. This intellectual heritage may be summarised as 'wilderness philosophy' (for discussion, see Oelschlaeger, 1991). Wilderness philosophy may be characterised as a set of nineteenth- and twentieth-century European-American practices and texts (most influentially articulated by Henry David Thoreau, John Muir and Aldo Leopold) that seek to encourage a spiritual and practical reverence for landscapes that are untouched by human hand and replete with the silent wisdom of authentic (as opposed to Westernised or 'degenerate') indigenous peoples. Any sense of contradiction between these two romantic contentions is made to disappear through the 'naturalisation' of Native Americans. The romantic archetypes of Native Americans – the ancient and taciturn Indian sage, the hunter, mimicking and silently tracking wild fauna – depict them as being as primitive, unchanging and spiritually potent as 'untouched' Nature itself. Represented without human voice or agency, Native Americans were portrayed as being fully part of the 'wilderness experience'.

These themes, and their translation into a desire to retreat into, or search out, wilderness are perhaps best exemplified in the nineteenth century within the work and life of Henry Thoreau. From early writings such as 'A Winter Walk' (1906; first published 1843) to the posthumously published *Maine Woods* (1987; first published 1864), Thoreau sought to experience primeval, untamed and forever untameable *Nature* … Man was not to be associated with it. It was Matter, vast, terrific, – not his Mother Earth that we have heard of, not for him to tread on, or be buried in … It was a place for heathenism and superstitious rites, – to be inhabited by men nearer of kin to the rocks and wild animals than we. (1987, pp.93–4)

Thoreau's wish to connect with 'the *solid* earth! … the *common sense!*' (1987, p.95) and 'Indian wisdom' (1906, p.131) led him to construct a rough home at Walden Pond in Massachusetts and live there in solitude between 1845 and 1847. In his account of his time at Walden (1962; first published 1854), he explains his desire to 'live so sturdily and Spartan-like as to put to rout all that was not life' (p.172). By pouring scorn on the 'luxuries', 'ornamentation', continuous 'dusting', 'Fashion' and 'effeminate' ways of urban America, Thoreau presents himself as an untameable and hardy frontiersman. Within uncultivated Nature, he implies, men may find an escape from the claustrophobic, alienated and artificial world of bourgeois/feminine society.

The connections between wilderness, 'primitive' peoples and masculinity that are apparent in Thoreau's work may also be seen to have fed into and drawn from the ideology of 'American westward expansion'. Tying these themes to the formation of a US national identity, the literary critic David Levernz (cited by Kimmel, 1995, p.118) explains that 'To be aggressive, rebellious, enraged, uncivilized; this is what the Frontier could do for the European clones on the East Coast, still in thrall to a foreign tyranny of manners.' Kimmel (1995) suggests that the demand for such 'fantasies of escape' intensified once the West had been conquered, and the prospect of 'real' escape from white civilisation disappeared. 'Western' novels, such as Crane's *The Red Badge of Courage* (1896), Wister's *The Virginian* (1902) and Wright's *When a Man's a Man* (1916) met a need for a vision of an unfeminised, unspoilt 'America': a 'purified, pristine domain … far from the feminizing, immigrant-infested cities, where voracious blacks and masculine women devoured white men's chances to demonstrate manhood' (Kimmel, 1995, p.129).

As this implies, the new men's movement is able to draw on a well-established and varied colonialist tradition of conflating ideologies of nature, gender and race. However, these themes have not simply been borrowed from the past by the 'new primitives'. Rather they have been reworked and reanimated to address the particular concerns and interests of contemporary middle-class men in the USA. As I have already indicated, one aspect of this process that interests me especially is the use of colonial myths to 'work through' and naturalise the movement's adherents' contradictory experiences of male authority. In order to open up this phenomenon to further scrutiny, I shall look at two forms of mythopoetic practice, retreats and initiation rites.

Before engaging with more typical mythopoetic praxis, it is important to note that a small number of mythopoetic retreats and initiation rituals attempt to reflect the pro-feminist minority current within the movement. Thus, for example, the 'Men's Rite of Passage' groups 'go into nature' (Wildwood, quoted by Wolf-Light, 1994, p.23) not in order to find a timeless source of masculine identity but to examine and deconstruct men's gendered identity. The 'beauty and rawness of nature', explains Wolf-Light (1994, p.23), 'work towards ... giving [a man] a larger context within which to see himself.'

However, deconstruction is not a leitmotif of the majority of retreat or ritual work. The commonest form of retreat involves a group of men camping, or staying in a cabin, for a few nights in a natural setting some distance away from the nearest built-up area. 'We want the men to leave the sounds and pressures of the city and come to a place with greater contact with birds, other animals, water, the earth, fire, and the rest of nature,' explains retreat organiser Shepherd Bliss (1992, p.96). The writer Aaron Kipnis (1992, p.163) offers an even bolder message to men: 'Enter the wilderness ... The Earth Father welcomes us, challenging us to be become stronger and deeper as men. We are at home in nature.'

Thus 'natural' landscapes are seen to be in harmony with 'natural' gender relations. As with Thoreau, the implied critique of the urban is also a critique of feminisation, which is, in turn, elided with modernity and change. In other words, the urge to get out of the city and plunge into the woods may be read as an attempt to escape from the unsettling, feminising environment of contemporary society; an environment that produces the 'soft male', the 'nice boy' rather than the 'hairy man' (Bly, 1990). Not unrelatedly, untouched, 'uncorrupted' countryside is also considered an appropriate place to act out cultural practices culled from the timeless and unchanging wisdom of tribal groups.

The retreat I will describe in detail has been called 'a classic' by Harding (1992: vi; for accounts of other retreats see Bliss, 1992; Pierson, 1992; Thompson, 1991). The First Southeastern Men's Conference, which lasted five days and was attended by 110 men, was, however, exceptional not only for its size and length but also for its importance. Held in 1989, the event brought together many of the most significant figures within the movement, including Robert Bly, Michael Meade and James Hillman. However, despite the conference's unusual status, the forms of social interaction seen at the conference, and its attendees' relationship to their physical setting, may be observed at a broad range of mythopoetic retreats.

The account of the event I shall draw on is provided by William Finger (1992) in an essay entitled 'Finding the door into the forest'. Finger, one of the event's organisers, begins his narrative by introducing us to the retreat's ability to pull men away from the urban and towards a more primal landscape:

> We came in car pools and airport shuttles, from 15 states and Canada, the majority from North Carolina and the D.C. area, to a secluded Blue Ridge Mountain church camp near Roanoke, Virginia. (p.100)

During the retreat, both this natural landscape and the naturalness of tribal cultural forms are repeatedly invoked as affirmations of masculinism. However, on close inspection, it becomes apparent that they are invoked, not *simply* to make powerless men feel powerful but to enable the attendees to use the event to 'act out' (i.e., to stage as a collective performance) and naturalise their contradictory experiences of power. Primitivist discourses are used to displace the men's fears and responsibilities surrounding gender roles away from the social or political realm and onto the terrain of Nature. In this latter landscape, men's power and men's essential character are seen as eternal and unchangeable. At the same time, the distinct and immutable character and power of the feminine (both within and outside male society) is 'recognised and respected'.

I shall provide two examples of the process of acting out powerfulness and powerlessness from the Blue Ridge retreat before describing the processes of naturalisation and displacement. The first relates to the attendees' practice of identifying themselves by reference to a wild animal (a Native American custom). Thus, soon after arriving at the site, the men divided themselves up into 'bears' ('the bulky and older guys gravitated to this phrase,' notes Finger), 'trout' (for those who 'dwell in pools of grief'), 'falcons' (for the 'fieriest') and 'dragons' (who organised and ran the event). Like many other activities undertaken during the five days, these primitivist archetypes were used to 'stage' experiences of domination and submission, of activity and passivity. Usually, this process was formalised. Thus, for example, the different animals were split up, assigned different tasks and different talking positions. During one discussion Finger (a dragon) recalls that he rose to speak 'even though Michael [the session leader] had asked to hear from the trout.'

> 'I'm one of the Men Who Feed the Dragons, Michael,' I said. 'You really didn't say we could speak yet.'
> 'Dragons can talk anytime,' he responded quickly (p.103)

Aside from such formalised power plays, Finger describes one 'unscripted' incident:

> The first night of dinner was ending, three falcons leapt from the edge of the dining hall … With red cloth flying from their arms and legs (each clan had its color), they mocked at the trout, mocked a peace banner, and as suddenly as they had struck, swooped back into the night. (p.101)

The power dynamics of this little drama find echoes in my second example of acting out powerfulness and powerlessness. This relates to the domination of the retreat by a few 'elders'. Bly, Meades and Hillman preside over the week like caricatures of tribal chieftains. They are wise and dignified; harsh yet fêted and submitted to. Bly's entry to the retreat was greeted by attendees forming their bodies into 'a snakelike corridor' whilst clapping 'a rhythmic welcome'. Bly passed through this applauding tunnel, observes Finger, and 'took his place beside Meade' (p.104).

Thus the retreat can be read as a performance; a performance in which 'primitive' ideas and cultural forms are used to express and re-enact conflicting experiences of power. These discourses are also used to naturalise these experiences and to displace political and social engagement. The attendees' repeated allusions to tribal and wilderness 'ways and wisdom' are invariably designed to lead beyond, or leave behind, everyday concerns and enable the men to reach the dynamic yet timeless male spirit. In the following account by Finger, Michael Meade leads attendees into the heart of maleness through the primal energies and rhythms of African drumming:

> Meade led us through a ritual adapted from an African tribe ... We got the hang of the odd-sounding chant and the accordion-like flow and moved into a feverish pace, impulses overriding reasoning ... Back and forth we went until we could barely speak, hoarse and exhausted. When we ended, I was transported into the timelessness of the mountains and birds and trees and ancient men. (p.102)

This imagined landscape is not simply a place where 'men rule'. It is a less crude but perhaps more insidious destination. For the mythopoetic movement is focused upon a space, a time, where men's and women's essential character is 'recognised', where the eternal struggle between feminine and masculine energies is 'understood'. It is to this comfortingly solid territory that men can return when 'we bring shame on our children and wives, then hate ourselves when we do' (p.105), and when 'we burst into sobs of grief, as men did throughout the week' (p.107). At the end of his article, Finger quotes Michel Meade: 'When you're in trouble, remember the dances' (p.108).

Thus the retreat, as its adherents are fond of saying, really does 'heal men'. But it does so not through clear-eyed self-reflection but through an act of displacement. As Kimmel (1995, p.139) writes,

> escape – from wives, partners, children, work, from adult responsibilities in general – has never provided the stable grounding for gender identity its promoters have promised ... the respite has only ever been temporary, and either must be constantly renewed in ever more bizarre ritual appropriations, or they lapse into the same politics of resentment and exclusion of anti-feminism and racism.

Despite the acuteness of this observation, there is a certain conservatism within Kimmel's censure. He implies that to be 'adult' means that men should avoid 'bizarre' rituals and buckle down to their 'responsibilities'. One is reminded that, for many critics, part of the transgression of the mythopoetic men is that they are 'ridiculous', that they do things that are not 'appropriate' for adults. The radicalism of primitivism has always relied on shocking those who consider themselves too grown-up to indulge in 'play'. It is a reaction that overlooks the fact that the issue of what being an adult male means in modern society is a central concern of the mythopoetic movement. Indeed, in *The Sibling Society*, Bly (1996) extends his earlier mythopoetic analysis through an attack on the infantilisation of contemporary culture. He suggests that it is the narcissistic and individualistic nature of consumer capitalism that subverts the possibility of growing up. In this context, Bly contends, rituals, especially those involving the symbolic passing of boys into men, have a role in demarcating the onset of responsible, socially aware, adulthood.

As with the wilderness retreat, the mythopoetic men's movement's use of ritual initiation evidences a keen interest in the power of Nature and tribal identity. However, it also provides some absorbing examples of another important characteristic of the movement's work, namely the mutative, playful nature of mythopoetic cultural appropriation. Initiation is widely seen within the movement as necessary for the emotional formation of young men and boys. As with so many other aspects of mythopoetic activity, the initiation rites that have been created to fulfil this need draw heavily on supposedly tribal, 'primitive' ideas and forms. Robert Moore (1992, p.9), one the 'elders' of the scene, explains that 'We lost all the technology of creating men out of boys when colonialism wiped out tribal cultures.' 'What people have to understand,' he continues,

> is that it wasn't just medicines that we lost when we failed to steward these Indian and other indigenous peoples' cultures. What they did not realize was that we also lost all the spiritual and psychological understanding. That understanding was of the radical necessity for initiation and empowerment according to gender … Male initiation was a much more rigorous form than the female because young males are more dangerous to the human community especially when the aggression in the lives comes on line biologically.

The specific examples of initiation I shall focus on are derived from Bernard Weiner's (1992) book *Boy into Man: A Fathers' Guide to Initiation of Teenage Sons* (Figure 4.4; for other examples, and discussion, see Daly, 1992b; Bly, 1991; Wiedemann, 1991). This text provides an account of one particular group of Californian fathers' attempts to initiate their sons. In summary, the book describes how the boys are taken off into the woods by their fathers for a weekend of symbolic ceremonies. The ceremonies are designed to be both enjoyable and serious. They include passing through human tunnels made by their fathers; having their feet washed and their foreheads anointed by their fathers; watching and listening as their fathers perform a specially scripted play on gender and adulthood; and, finally, receiving an 'official welcome into the inner circle of manhood' (p.24).

As with the retreat described earlier, the gender essentialism of these performances is articulated through constant allusions of the sagacity of 'primitive' peoples and landscapes. In the play performed for the boys, it is explained that 'Tonight, we are incorporating some of our own history, and that of other men in various cultures, to honour and celebrate your entry into the world of men' (p.32). At another point in the ceremony the boys are instructed that 'To help you connect with the deepest aspects of yourself, and your relation to the world and this higher level, tonight you will stay in the woods by yourself, at the sacred spot you have selected' (p.35).

However, it is apparent from Weiner's account that mythopoetic cultural appropriation is neither simple nor entirely unsubversive. In particular, his account reveals three aspects of the appropriative process that distinguish it from a naive repetition of colonialist fantasies. These are (1) adherents' interest in creatively mixing up cultural forms and traditions; (2) adherents' willingness to rework creatively 'old' and 'new' ideas and symbols; and (3) adherents' ability to play, to actively enjoy the process of cultural reinscription. Thus rather than simply copying colonialist myths, the mythopoetic movement imaginatively reworks them.

Figure 4.4 *Boy into Man: A Fathers' Guide to Initiation of Teenage Sons,*
B. Weiner (1992) (front cover)

As this implies, there is no pretence by the fathers that the tribal and
folkloric masks and performances they stage are authentic. 'We rounded
up various props, costumes, masks,' explains Weiner (p.15). The fathers
are more interested in *ad hoc* bricolage than cultural mimesis. Some of the
assembled objects and recited texts appear modern, some ancient, some
tribal. Sometimes they appear to be all of these things together. The ceremonies, symbols and text devised for the weekend are not designed as
imitations but as objects and actions with multiple, unspecific, heritages.
Despite the essentialist discourses that cohere mythopoetic ideology, the

movement's performers are no respecters of cultural purity, whether 'tribal' or Western.

The ceremony which begins the weekend may be taken as an example of the three mutative processes identified above. At this time the boys are still in school. Reworking a 'tribal' custom, the fathers decide that the boys should be symbolically kidnapped 'from the security of their school-womb' (p.17). Their pleasingly silly method of achieving this aim is described by Weiner:

> I had located some giant, four-foot-high hands – which had been used in a play that I'd seen 15 years previous – and two of the men got inside the hands and practised "beckoning" the boys out. We began drumming, the gate leading to the schoolyard was opened, the hands beckoned, the fathers called out their son's name, and one by one, each boy exited the schoolyard and headed for the four cars that would drive them to the initiation site. (p.17)

The 'snatching' of the boys may be said to draw on primitivist discourses rooted in the colonial era. However, it also represents an imaginative reinscription of colonialist mythology, a reinscription that playfully mixes up cultural codes and expectations. In my conclusion, I shall be returning to this contentious, and underdebated, issue.

Who wants to be white any more? The appeal of primitivism outside the West

Years ago, as a first-year student in London, I had a Nigerian friend of Indian heritage who liked to dress up. In particular, he liked to wear what he called his 'tribal stuff', which consisted of cuffs and collars of animal fur and brightly patterned shirts, decorated with bones or masks. Dressing up 'like an African' was for him an enjoyable performance (I thought he looked pretty silly but since, in the name of fashion, I had shaved half my hair off and wore surgical bandages round my arms I was hardly in a position to accuse anyone of errors of taste). 'Tonight I'm dressing up like a total savage,' he told me one evening, sporting his loudest shirt, explaining 'Who wants to be white anymore?' It is a familiar primitivist sentiment. And it is not one necessarily employed only by whites. As previous chapters have shown, white identity has been disseminated across the world as an aesthetic norm and corporeal signifier of modernity. Everyone, whatever their skin colour, knows something of what it means to be white, more specifically the association of whiteness with the norms of 'modern society'. It follows that the notion of escaping whiteness into a more 'human' and less sophisticated realm, the realm of the primitive, is now available to many more people then merely those of European heritage.

Non-white, non-Western primitivism has tended to evade the scrutiny of scholars of both anti-colonialism and primitivism. It is an oversight that has been encouraged by the increasingly dated notion that the West alone is responsible for primitivism. As this section aims to show, primitivism and the escape from whiteness are now themes recognisable within many cultures around the world. I cannot pretend that the examples I have chosen to illustrate this process – namely, some themes in Sri Lankan cultural nationalism; issues raised by a festival of masks in Côte d'Ivoire; and the views of Senghor and Fanon on negritude – are either representative or offered here in any great depth. However, they should serve to flag up some of the problematics of non-Western primitivism and, I hope, to provoke readers to look further into what, at present, is a neglected area of enquiry.

The global reach of the ideologies and practices of modernisation means that there are now no parts of the world where social and economic life is not constituted, in some way or another, in relation to modernity. Often this process has entailed the adoption and adaptation of notions of the modern as an ideal, a 'developed' condition to be admired, constantly striven for and, where possible, sustained. However, any society constituted in relation to modernity is also constituted in relation to pre- and/or non-modernity, to notions of what is outside or threatening to the modern. Given both the racialised nature of modernity and the influence of a wide spectrum of Western ideas across the world (*including discourse of resistance*, for example, romanticism, primitivism, youth rebellion, feminism, Marxism and anarchism), it is not surprising that attempts to subvert, or merely critique, modernity in non-Western countries have often been articulated in terms that appear to echo Western debates and Western ideologies. It should also be noted, however, that the notion that non-Western culture echoes Western concerns needs to be approached with some care. Both the particularity of influences and the range of influences at work within any one society require attention. Even more importantly, the fact that discourses that can be claimed as Western, such as 'pastoral romanticism', or 'primitivist nostalgia', may also be able to be categorised in other ways, ways that suggest that their associated practices and ideologies are 'indigenous' and do not derive from the West, at least not in any simple, mimetic fashion.

Spencer's (1995) studies of occidentalism in nationalist discourse in Sri Lanka provide illustrations of a number of these points. In Spencer's account, connections between anti-colonial radicalism and Western romanticism emerge particularly clearly in the writings of 'ruralist' Sri Lankan intellectuals. I shall take one of the most prominent figures in this latter group as an example, the art critic Ananda Coomaraswamy. Writing in the late nineteenth and early twentieth centuries, Coomaraswamy looked forward to a Sri Lanka freed from the alienated world of capitalist Western domination. His pastoralist, utopian vision suggested that there existed in Sri Lanka a more organic, more authentic, past, a legacy that was being corrupted by, and needed to be rescued from, Western influence. For Coomaraswamy, it was in Sri Lankan village life and the conditions of 'medieval' cultural production that one could find the soul, the authentic essence of the country. Only by returning the country to these roots could it become truly independent and free. And yet this construct of Sri Lanka's heritage and political future exhibits Coomaraswamy's

English influences as much as his native inspiration. Noting the influence of pastoralist notions of village life in the East developed by British anthropologists of the period, Spencer (p.247) writes that 'In Sri Lanka the village community of Victorian social theory lives on as the *fons et origo* of post-colonial political programmes.' Another English influence upon Coomaraswamy was the pastoralist anarchist William Morris. Indeed, Coomaraswamy's most significant book, *Medieval Sinhalese Art* (1956; first published 1908) was printed at Morris's Kelmscott Press. This confluence of influences did not escape Coomaraswamy's attention. Indeed, he looked upon it as indicating the potential for a universal movement away from Western modernity and towards a more natural social order. Writing of his use of the Kelmscott Press, he explained that it was 'an illustration of the way in which the East and the West may together be united in an endeavour to restore that true Art of Living which has for so long been neglected by humanity' (cited by Spencer, 1995, p.245).

As this example implies, the positioning of 'Third World' national and pan-national projects as forms of resistance to the alienated, instrumental logic of modernity has tended to rely on, and sustain, the conflation of the terms 'Western' and 'modern'. The evocation of an unspoilt, pre-modern, non-Western national essence may be seen as a necessary fiction for post-colonial societies attempting to claim a distinctive national and racial project. Other factors that may encourage the expression of primitivism in non-Western societies include the nature and impact of tourism and ethnicised political struggles over resources. With respect to the latter, it is pertinent to note that claims to the possession of an authentically pre-modern traditional or tribal identity have been mobilised by various communities around the world in order to win both international attention and grants and to secure national political representation. Communities favoured in this manner tend to be numerically unthreatening, are able to be cast as 'maintaining customs' and are relatively unspoilt by contact with the 'outside world'. They are also sometimes classified as 'indigenous', a word that has come to be accorded considerable political and ethnic resonance (see Kemf, 1993). In many countries, the claims to be following a simpler and more natural way of life associated with 'indigenous' communities have been incorporated into national and nationalistic discourses. The best-known examples of this process are in Latin America, where nation-building projects have often drawn on ideologies of indigenism (see Knight, 1990).

Latin America also has many examples of the deployment of primitivism in the service of tourism. The desire to come into contact with the exotic, untouched places and people of the continent may be traced back to the romanticism that inflected the founding and European exploration of Latin America in the nineteenth century. However, in the mid- and late twentieth century this type of representation has become integrated into the economies of many of the poorest areas in the region. For national and regional governments with jurisdiction over such territories as the Peruvian altiplano or the southern states of Mexico, the need to project these areas' non-European, 'uncorrupted' heritage as a lure for tourists has come to be seen as an important ingredient in their economic development. This form of tourism is reliant on a number of markets. Both Western tourists and travellers from the metropolitan regions of Latin America are encouraged to experience the excitement of voyaging into more primitive realms. Perhaps, though, the people who understand the thrill of primitive spectacle better than anyone are those who are considered to be producing it. For it seems that, in many contemporary tourist encounters, those who were previously merely the human objects of the primitivist gaze have become engaged in sustaining primitivist conceits. The self-conscious primitive tries to gauge what the tourists want and finds him- or herself economically compelled to provide it. This may not be an entirely new phenomenon, but it has become sufficiently ubiquitous today for the suspicion of 'fakes', the search for the 'real thing', to become a dominant factor in the arrangement of tourist itineraries and interactions with 'the natives'. The tourist demands to see something resembling her or his own pre-formed image of 'the primitive', but the latter must not be too self-conscious in his or her role, must not appear to be trying too hard to get it right, otherwise the performance, the promise of escape, will end in disappointment. By the turn of the twentieth century, it is not unknown for wearied world travellers to bypass the advertised tour to the 'remote hill villages', the stop-off at the 'tribal' hut selling ethnic 'objects': primitives that fail to get it 'right' can quickly find themselves shunned by Western authenticity hunters. I can think of no more concise, or cruel, commentary on the latter process than that offered by an acquaintance of mine after returning from year travelling the world. Most local traders in the tourist market, he opined, 'just seem like a bunch of poor people selling junk that wasn't even fashionable ten years ago. They don't have a clue what we really want.'

An interesting account of the development of a knowing, self-conscious primitivism in the service of both tourism and the nation-building process may be found in Steiner's (1997) depiction of the staging of the national festival of masks in Côte d'Ivoire. As Steiner explains,

> Following the dramatic collapse of the price of cacao and coffee in the world market ... politicians scrambled to find not only a new source of foreign income but also a new gathering point for nationalist sentiment. The mask was thought to be capable of achieving both. On the one hand, it fuelled the Western imagination through its mystery and exotic appeal. On the other hand, it reconciled growing ethnic divisions by elevating the symbolism of the mask – with its plethora of ethnic styles and interpretations – to a single, national icon. (p.673)

The tourist brochure for the festival of masks held in 1983 promised that it would offer 'the equivalent of Carnival in Rio, with an added element of the profound soul and mystery of "non-commercialised" Africa' (cited by Steiner, 1997, p.673). In fact, the festivals were widely judged to have been unsuccessful. The expected crowds did not arrive. The attempt to blatantly rip the iconography of masks from its local setting and employ it to conjure a homogenised national image was resented by people within the country and seen by tourists as inauthentic and 'too staged'.

Claiming *for oneself* the status of being authentic, of being primitive, is always problematic. For once one begins to describe oneself in such a manner one is likely to be judged to be too sophisticated, too knowing, to be a genuine example. Unless carefully managed, any claim to be primitive can appear like a disingenuous and manipulative performance. In this sense, a primitive identity cannot be asserted without falling into contradiction: to aspire to it rules one out of ever obtaining it. Nevertheless, primitivism exists as a current, sometimes minor, sometimes predominant, within many self-consciously anti-Eurocentric expressions of non-Western identity. The literary and artistic movement known as negritude – tellingly described by Fanon (1986, p.135) as 'this unhappy romanticism' – may be taken as an example. Although it would be a gross misrepresentation to reduce negritude to primitivism, the characteristic themes of primitivism – an emphasis on Nature over culture, authenticity over alienation, an attraction to the simple over the complex – surfaced repeatedly in its work. Negritude was the product of black intellectuals based in Paris in the 1930s. Although its members were diverse in their output, the movement is associated with the attempt to express a modern black sensibility, a sensibility able to explore the meaning and value of Africanness in the twentieth century. The negritude writers influenced and were, in turn, influenced by the white avant-garde groups then current in Paris, especially surrealism. Indeed, André Breton wrote some of the first appreciations of the negritude poet, Aimé Cesaire. Expressing his own thoughts on the intimacy of the links between the new art and the explorations of the negritude writers, another poet (and later president of Senegal), Léopold Senghor, noted that 'the African aesthetic is "that of contemporary art"' (Senghor, 1997, p.634). Senghor's remark seems to locate a happy confluence of traditions, a place where the white avant-garde and black anti-colonial intellectuals could meet and define a shared identity. However, Senghor's comment evades the fact that 'the African aesthetic' was prized in Europe precisely because it appeared to be the product of the primal creative energies of *tradition-bound and Nature-bound* people. The aesthetic of 'contemporary art' was prized for the opposite reason. It was valued because it appeared to be the product of sophisticated, cosmopolitan people who had chosen to defy convention. Not surprisingly, it was the latter arena, not the former, that sustained the negritude group. Their aesthetic was designed for a highly self-conscious, 'Western-educated', audience. Thus, when in his defence of negritude – 'Negritude: a humanism of the twentieth century' (composed in French like all his other literary works) – Senghor offered a vision of 'African thought' as holistic, spiritual and organic, he appeared to be constructing and deploying primal Africa in much the same way as the European avant-garde. 'Since the Renaissance,' he wrote, 'the values of European civilisation had rested essentially on discursive reason and facts, on logic and matter' (p.630). By contrast, 'for the African' a mystic holism prevails: 'living according to the moral law means living according to his nature ... Thus, he transcends the contradictions of the elements and works towards making the life forces complementary to one another' (p.633). Senghor's vision offers a homogenising, mystified view of 'the

African mind', a view that situates Senghor less as a mouthpiece for authenticity and more as its gatekeeper or arbiter.

A critic of negritude, the Martiniquan activist Frantz Fanon, provides a more politically complex account of the desire, the need, for non-whites to establish racial identity and meaning in a white-dominated world. A clinical psychologist by profession, Fanon was particularly alert to the psychological damage of white racism. In *Black Skin, White Masks* (1986; first published 1952) he sought to analyse the desire for white ness felt by the black person in a 'white world'. 'Out of the blackest part of my soul, across the zebra striping of my mind, surges this desire to be suddenly *white* ... I wish to be acknowledged not as *black* but as *white* ... I marry white culture, white beauty, white whiteness' (p.63). However, another current is also clearly evident in *Black Skin, White Masks*, namely a desire to explore the way that non-white racial identity is a fabrication, a conceit developed in the context of, and in opposition to, myths of white identity. 'I shall demonstrate,' wrote Fanon, 'that what is often called the black soul is a white man's artefact' (p.16). More pointedly, he cites negritude as an example of racial myth making, a form of essentialist delusion that remains trapped in the categories and conceits of white racism. Yet Fanon also suggests that it is equally delusional to imagine that oppressed people can gain any kind of agency over their lives without such myths of identity. The costs, he implies, of withdrawing from the myth-making process are simply too high. In order to have a place from which to resist the crushing weight of the lie of whiteness, African-heritage people must construct a lie of blackness. The following passage captures Fanon's attempts to wrestle with these problems. He starts with a somewhat mocking tone, citing Senghor and a friend from the United States with a similarly essentialist belief that negroes possess qualities of 'emotion', whilst whites possess qualities of 'reason'.

> From the opposite end of the white world a magical culture was hailing me. Negro sculpture! I began to flush with pride. Was this our salvation? ... Black Magic, primitive mentality, animism, animal eroticism, it all floods over me ... Yes, we are – we Negroes – backward, simple, free in our behaviour. That is because for us the body is not something opposed to what you call the mind. We are in the world. And long live the couple, man and Earth! ... Listen *'Emotion is completely Negro as reason is Greek'* ... I made myself the poet of the world ... The soul of the white man was corrupted, and, as I was told by a friend who was a teacher in the United States, 'The presence of the Negroes beside the whites is in a way an insurance policy on humanness. When the whites feel that they have become too mechanized, they turn to the men of color and ask them for a little human sustenance'. At last I had been recognised, I was no longer a zero. (pp.123–9)

Fanon's initially derisive manner is sustained as he recounts how primitivism is itself a conceit of the West, that it provides no secure foundation for an unchanging, pure, African identity: 'I tested the limits of my essence; beyond all doubt there was not much left of it' (p.130). He goes on to cite Sartre's placing of negritude 'as the minor term of a dialectical progression: the theoretical and practical assertion of the supremacy of the white man is its thesis' (Sartre cited by Fanon, p.133). However, for Fanon, Sartre's contextualisation of negritude also reads as a kind of possession, a form of colonial knowledge of the other made from the safety of an unthreatened, ever dominant, white vantage point. Leaving aside his earlier mocking tone, he argues that all this prising apart of blackness, its placing and deconstruction, destroys any possibility of resistance to white dominance. Fanon does not, it seems, disagree with Sartre theoretically, but on a political and emotional level he cannot tolerate his blackness being dismissed as a 'minor term in a racial dialectic'. I will conclude this section with what sounds to me like a cry of despair, a declaration that he, Frantz Fanon, must, despite everything, be a negro, be the other of the white, or be destroyed.

> I defined myself as an absolute intensity of beginning. So I took up my negritude, and with tears in my eyes I put its machinery together again. What had been broken to pieces was rebuilt, reconstructed by the intuitive lianas of my hands. My cry grew violent: I am a Negro, I am a Negro, I am a Negro.... (p.138)

Conclusions and a remark on 'cultural appropriation'

Primitivism has often been treated as a rather eccentric and minor example of social fantasy. Its practice has been associated with a few individuals, odd creatures who, for all their bombast, are fundamentally ridiculous. Indeed, if any one characteristic weaves its way through existing commentaries on primitivism it is their condescending tone. Primitivists are depicted as deluded, as racists and as immature. Such portraits imply, of course, that 'we' are the opposite of all these things, that 'we' have no need for primitivism. Yet if we accept that modernity was and is racialised, that it is Eurocentric, then primitivism – as the voice of pre- or non-modernity – starts to appear as an immanent and inevitable moment of critique, a site of resistance to modernity produced within modernity. That this site of resistance has often also been a site of racist fantasy is a dilemma faced by all those who wish to confront and redefine the meaning of the modern.

The four examples I have introduced in this chapter provide distinct perspectives on the problem of primitivism. The first three illustrations appeared to conform to what is now a familiar theme in modern culture, namely 'white escape'. However, as the fourth and final section, on non-white primitivism, indicates, the flight from whiteness is not necessarily enacted or conceptualised by white people. This intervention may usefully cause us to start to question the whiteness of my other examples. Avant-garde primitivism, the retreat into nativism and wilderness, the search for myths of unchanging masculinity within tribal cultures: in what sense are these white activities or, indeed, necessarily performed predominantly by white people? It is a provocative enquiry precisely because it is taken for granted in so many quarters that primitivism is rooted in European creativity, in white agency. Primitivism may be laughed at or even seen as reactionary, but if it is construed as uniquely European it is the political will, the ability to act against modernity, of everybody else on the planet, that is undermined and marginalised. It is not just whites who have the right to be 'ridiculous'.

I want to conclude this chapter by returning to another controversial aspect of primitivist practice. I am referring to the issue of 'cultural appropriation' (see also Fung, 1993). Primitivism has been widely read as 'stealing other people's' cultures. It is accused of being 'inauthentic'. Thus, for example, in *The Imaginary Indian*, Francis (1992, p.109) denounces Grey Owl as a 'plastic shaman', a fraud who obscured the real voice of real Native Canadians. Indeed, Francis goes on to claim that the recent political advances of First Nation peoples in Canada have been achieved through a rejection of such 'inauthentic' representations. Native Canadians now 'enjoy unprecedented political power' and are recognised 'as one of the founding peoples of Canada,' he writes, adding that 'All this came about because Native people refused to live within the stereotypes White people fashioned for them' (p.220). Kimmel and Kaufman (1994) offer a similarly morally indignant condemnation of the mythopoetic men's movement:

> can [non-Western] rituals be ripped from their larger cultural contexts, or are they not deeply embedded in the cultures of which they are a part[?] … To see a group of middle-class white men appropriating 'Indian' rituals, wearing 'war paint', drumming and chanting, and taking on totemic animal names is more than silly play … It is politically objectionable. (pp.272–5)

It might be argued that, if the mythopoetic movement's practices are based on racist colonial fantasies, then just whose culture are they appropriating? Can one appropriate one's 'own' inventions? The implication of these types of attack is that white middle-class men should adhere to their 'own' culture. Yet what is this culture? And who is permitted to define it? Perhaps, on some issues at least, the 'new primitives' are less essentialist than their critics. When set against innumerable incidents of white people stealing and profiting from non-white societies, this line of argument may appear glib. Yet it does raise an important point about the uses of primitivism and its transgressive qualities. I am not convinced by Francis's assurance that Native Canadians have achieved status only by rejecting 'white stereotypes' and by being 'authentic'. Indeed, ironically, the notion that non-white peoples do not participate in primitivism, do not use and deploy it to their own ends, implies that they are so remote from modernity, so unspoilt by the dominant discourse of the twentieth century, as to be *truly* exotic. Moreover, I think it must be admitted that primitivism necessarily attempts, at some level, to cross the boundaries of cultural and racial purity. It usually combines this anti-racist gesture with racist stereotype. But its socially subversive qualities need to be monitored just as closely as its conservative ones.

1.   Stelarc is just one artist amongst a generation of cultural workers who, in the 1980s and 1990s, have sought to use their bodies as sites upon which to play with themes of identity and otherness (see Bonnett, 1994). Other examples include the French artist Orlan (amongst her various body modifications Orlan has sculptured flesh horns into her head) and Fakir Musafar (who has performed numerous tribal piercing ceremonies).

2.   Grey Owl was not the first non-native person to fabricate a Native American identity. Buffalo Child Long Lance, whose autobiography, *Long Lance*, appeared in 1928, claimed to be a Cherokee from North Carolina. He was, in fact, Sylvester Long, who as Francis (1992, p.127) notes was 'of mixed Indian, White and probably Black ancestry. Whatever their actual background, the [Long family] were known … as "colored"'. Francis suggests that, for Long, the adoption of a native identity was a form of escape from the racist oppression of the South.

3.   *Two Little Savages* may be regarded as a founding text for Seton's Woodcraft League, a precursor to the Boy Scouts that also acted as a forum for Seton's socialist and 'pro-Red Indian' politics. In his 1936 book *The Gospel of the Redman*, and with the Great Depression very much in mind, Seton (cited by Francis, 1995, p.157) wrote that 'The Civilization of the Whiteman is a failure, it is visibly crumbling around us … [Indians are] the most heroic race the world has ever seen, the most physically perfect race the world has ever seen, the most spiritual Civilization the world has ever seen.'

CHAPTER 5

White identities and anti-racism

Introduction

> It would be hard to imagine someone writing a book about what it
> means to be white. Most white people don't consider themselves to be
> part of a race that needs examining. They are the natural order of
> things. (Saynor, 1995)

Since James Saynor offered these comments at least a dozen books about
whiteness written by white people have appeared. In fact, quite a few
existed at the time Saynor claimed that such a thing would be 'hard to
imagine'. It is, of course, always tempting to assert that one's particular
area of enquiry is novel, that it is under-discussed. In what has become a
familiar gesture, Saynor links the idea that whites' identities have rarely
been written about with the contention that whites are 'invisible' in racial
discourse (see also Dyer, 1997; Roediger, 1992, 1994; Ignatiev, 1995; Frankenberg, 1993). It follows that whites must be 'outed', that they must be
dragged from the shadows. Whilst not disagreeing with this prescription
I would suggest that such claims need to be treated with some care. We
should not forget that until relatively recently the attributes of the 'white
races' were not a subject about which white people were known to be
particularly reticent. Colonial and racist anthropologies and histories produced a voluminous literature on the superiority of white civilisation.
Anti-racists have also contributed to the subject for several decades, a
considerable body of work existing on the reproduction, identification
and overturning of white racism (for example, Katz, 1978; Wellman,
1977). Since white identities have been subjected to such examinations,
we need to ask why contemporary analysts appear convinced that whiteness is invisible. There are a number of answers to this question, but the
most pertinent turns on the issue of *how* whiteness has traditionally been
addressed. For, until recently, racist and anti-racist commentators – most
especially but not exclusively white commentators – have shared a tendency to naturalise whiteness, to treat it as an ahistorical and geographically undifferentiated racial norm (for some exceptions to this pattern see
Roediger, 1998a). In other words, it is the fact that whiteness has been

approached as, to use Saynor's words, 'the natural order of things' that has structured its representation. A naturalised, normative identity is not necessarily an undiscussed identity. However, the discussion it supports is of a very particular kind. It is inevitably grossly generalising, it inevitably reifies its subject matter, and it is – perhaps not inevitably, but probably – prone to cliché, the repetition of commonsense knowledge and the banality of polemic. Thus the *meaning and formation* of whiteness are taken for granted; *the history and geography of the subject made invisible.* Now this sorry list is hardly a problem if we wish to produce white supremacist knowledge. Indeed, white racists would wish to defend and sustain precisely such an 'examination'. If whiteness is accepted as a static, homogeneous 'thing', such people would surely welcome it being 'outed'. However, for anti-racists these forms of essentialism are a problem, one that needs identifying and challenging (see also Bonnett, 1999).

In this chapter, I will show that the reification of whiteness is a central current within English-language anti-racist thought and practice. I shall, in addition, be suggesting that this process has deleterious consequences for the anti-racist project. More specifically, I argue that the reification of whiteness has enabled white people to imagine that their identity is stable and immutable and, relatedly, to remain unengaged with the anti-racist historisation (and denaturalisation) of racial meaning.

I shall be focusing on British and North American anti-racist debate. The active and diverse anti-racist traditions within these societies are currently experiencing a series of intellectual and practical crises and opportunities. Traditional anti-racist paradigms are being challenged by a variety of forces, including a so-called 'new ethnic assertiveness' (Modood, 1990a), conservative anti-anti-racism and postmodern critical interventions (for discussion, see Bonnett, 1993a; Rattansi, 1992). Ali Rattansi (1992, pp.52–3) has argued that 'if anti-racism is to be effective' it will be 'necessary to take a hard and perhaps painful look at the terms under which [it has] operated so far.' More specifically, Rattansi calls attention to anti-racists' inadequate and simplistic modes of racial representation. Unprepared to acknowledge the 'contradictions, inconsistencies and ambivalences' (p.73) within white and non-white identities, orthodox anti-racism appears ill-equipped to engage creatively with the fluid and complex forces of the racialisation process.

My account commences with some instances of the reified nature of whiteness within contemporary British and North American anti-racism. Three types of example are offered: the term 'white' in anti-racist nomenclature; whiteness and the 'new ethnic assertiveness'; and essentialist interpretations of the attributes of white identity. Having introduced the way anti-racism has erased and reified whiteness I turn to review recent anti-racist activism and writings that appear to offer an alternative to these discourses. Four forms of 'white studies' are addressed: (1) anti-racism in white areas; (2) the literature of 'white confession'; (3) engaging 'excluded whites'; and (4) historical geographies of whiteness. As we shall see, although all of these forms, to a greater or lesser extent, articulate problematic notions of whiteness, the historical geography approach does point towards the possibility of more nuanced, and strategically and theoretically more useful, anti-racist readings of whiteness. The chapter concludes with an assessment of the practical and theoretical implications of anti-essentialist readings of whiteness.

Anti-racism and the reification of whiteness

It is worth repeating that I am not claiming in this chapter that the supposed attributes of white racial consciousness have been overlooked by anti-racists. What I am suggesting is that whiteness has tended to be approached by anti-racists as a fixed, asocial category rather than a changeable social construction. In other words, anti-racists have, for the most part, yet to become aware of, and escape from, the practice of treating whiteness as a static, ahistorical, aspatial, 'thing': something set outside social change, something central and permanent, something that defines the 'other' but is not itself subject to others' definitions.

It is also my contention that reifying myths of whiteness subverts the anti-racist struggle. They create an essentialising dynamic at the heart of a project that is necessarily critical, not only of racial stereotypes but also of the 'race' concept itself. They also lead towards the positioning (or self-positioning) of white people as fundamentally outside, and untouched by, the contemporary controversies of racial identity politics. For within much contemporary anti-racist debate whiteness is addressed as an unproblematic category (albeit with negative attributes), a category which is not subject to the constant processes of challenge and change that have characterised the history of other ethnic and racial names. This process enables white people to occupy a privileged location in anti-racist debate; they are allowed the luxury of being passive observers, of being altruistically motivated, of knowing that 'their' racial identity might be reviled and lambasted but never actually made slippery, torn open, or, indeed, abolished.

To exemplify these points, I will offer three instances of the erasure and objectification of whiteness within anti-racism.

'White' in anti-racist terminology

Anti-racists have often sought to show that language matters. Writers as diverse as Gilroy (1987) and Modood (1988) have insisted that racial terms are neither politically neutral nor static, that they have contested histories. Not unrelatedly, anti-racists have tried to encourage the use of racial nomenclature that embodies political reflexivity and discourage expressions that appear imposed, outmoded, offensive and/or phenotypically reductive.

However, there is one exception to this linguistic sensitivity. For 'white' tends to be excluded from anti-racists' list of acceptable or debatable racial nouns and adjectives (see, for example, ILEA, 1983; Gaine; 1987). Indeed, in the majority of anti-racist work the meaning of whiteness appears to be considered beyond dispute, its boundaries obvious. Thus the reification of whiteness is enacted as an erasure: whiteness is simply left out.

For a typical example we may turn to the anti-racist policy documents published by the Inner London Education Authority (ILEA) in 1983. The ILEA's policies were highly influential for a few years, both in the UK and elsewhere. Before the authority was abolished by central government (partly because its anti-racist commitments were equated with political radicalism by the Conservative government) it offered a clear lead to other metropolitan authorities on matters of race equality. The ILEA's approach to race hinged on the concept of whiteness. Its policy documents make liberal use of the term. Whites are positioned as key agents within anti-racism. They are instructed, cajoled, reprimanded. For anti-racism to exist, it is implied, it is whites who must be anti-racist. Thus, in *A Policy for Equality: Race* (1983) references are made to the fact that 'white people have very much to learn from the experiences of black people' (p.5) as well as to the way 'racism gives white people a false view of their own identity and history' (p.7). The ILEA, it seems, considered that 'a white community' existed in London, a distinct and obvious group of people who had their 'own identity and history'. The reification inherent in such interpretations is cemented into unassailable common sense by this particular document's 'Note on Terminology'. The note begins with the somewhat cryptic clause, 'The following terms are used:'. Beneath the authority's colon we find explanations of the meaning of three racial descriptions: 'Afro-Caribbean', 'Asian' and 'black'. There is no entry on 'white'. It is the only racial term mentioned in the main text whose meaning is not explained. Thus white is allowed to 'speak for itself'. It is permitted the privilege of having an obvious meaning, of being a normal rather than an exceptional case, of being a defining, not a defined, category.

Destabilising blackness/stabilising whiteness

Since the late 1970s the meaning and boundaries of blackness have been subject to considerable scrutiny in both North America and Britain. Within Canada and the USA the so-called ethnicisation of blackness has been associated with the development of new, hyphenated identities, most significantly 'African-Canadian' and 'African-American'. In Britain, the examination of blackness has been complicated by the term's diversity of applications. Drawing on the equation of the term with people of African heritage that exists in North America, there has been a parallel emergence of an 'Afro-British' (or, less commonly, 'African British') category. However, black in Britain is also often applied to 'Asians' (that is, people of South Asian heritage). Sometimes this usage is 'political', in the sense that 'black' is used as a collective noun for all those who experience racism. On other occasions, though, it simply reflects a cultural practice (especially common amongst whites) of lumping together everyone who is 'dark-skinned' as black. Despite this variety of usages, it is the employment of 'black' as a political label that has attracted most debate. This partly reflects the fact that the issue has been enacted as a controversy within anti-racism, and amongst anti-racists a political usage of 'black' has, until recently, been dominant. More specifically, a number of writers, including Tariq Modood (1988; 1990b) and Ali Rattansi (1992), have made an important intervention in British anti-racism by arguing that the anti-racist practice of defining South Asian people as black denies the multi-faceted nature of the cultural identities of this group. As this focus implies, it is only certain specific 'ethnic communities' that are being dis-embedded from the monoliths of orthodox anti-racism. The corollary of blackness – whiteness – has been left entirely undisturbed. Thus we find that, even within an article entitled 'Beyond racial dualism' (Modood, 1992), the mythologies of blackness are attacked whilst those of whiteness are left undiscussed. Only one-half of the dualism is surmounted; white-ness remains intact whilst blackness is demolished. This selectivity is, I would submit, a potentially disastrous facet of contemporary discourses that seek to reflect and/or assert the so-called 'new ethnic assertiveness' (Modood, 1990a). It is important not to forget that, within Britain, anti-racists' constructions of blackness were designed to serve an important purpose: to establish and support a 'community of resistance' within and against a white racist society. In other words, 'black' was a politically self-conscious category necessitated by the existence of a naturalised and unselfconscious 'whiteness'. The close links established between black-ness and a militant, relatively unethnicised identity may be seen to have made it incapable of incorporating and sustaining divergent ethnicities in the same way as whiteness appears to do. However, this vulnerability is rooted in the anti-essentialist tendencies inherent within any project that attempts to privilege political over natural solidarity. Thus, it is a some-what savage irony that it is 'black', a relatively sophisticated and 'self-knowing' construction, rather than 'white', the unselfcomprehending entity that forced 'black' into existence, that is on the receiving end of so many contemporary attempts to destabilise 'racial monoliths'.

'White racism' as an essentialist category

As I have implied, anti-racists have often placed a myth of whiteness at the centre of their discourse. This myth views 'being white' as an immu-table condition with clear and distinct moral attributes. These attributes often include being racist; not experiencing racism; being an oppressor; not experiencing oppression; silencing; not being silenced. People of colour are defined via their relation to this myth. They are defined, then, as 'non-whites', as people who are acted upon by whites, people whose identity is formed through their resistance to others' oppressive agency (this point is made well by Modood, 1990b).

To exemplify this point, I will turn to a Canadian anti-racist text. In 1994, a Toronto-based private anti-racist consultancy called the Doris Marshall Institute (1994) produced an address on 'Maintaining the ten-sions of anti-racist education' in the journal *Orbit*. About one-third of the piece is devoted to a commentary on the relationship between a white member of the group and the commissioning of the article. 'The editors of *Orbit*', notes the Institute (p.20), 'approached Barb Thomas, one of the core members of the Doris Marshall Institute (DMI), to write an article on our approach to anti-racism education.' However, this decision was in error, notes the Institute, because

> Barb is a white woman who does not experience racism. There is a mounting, legitimate critique of white people getting the space, resources and recognition for anti-racism work. (p.20)

The Doris Marshall Institute goes on to explain that the role of anti-racism is to strengthen 'the voices and leadership of persons of colour' and to

> insist that white people take responsibility for confronting racism and assist white people in this when necessary, and challenge speakers and writers to make explicit their voices and locations and what their limits and possibilities are. (pp.20–21)

It is being argued here that the 'limits and possibilities' of a racialised subject's engagement with anti-racism are established by her or his experiences of racism. Although this proposition has a certain superficial straightforwardness (though see Miles, 1989, and Fuss, 1989), it relies on a number of essentialist demarcations and categorisations. More specifically, the experiences of 'white people' are presented as manifest and unchanging. The characteristics of whiteness are removed from a social context and set outside history and geography. It is important to note that this process does not occur simply because racism is being associated by anti-racists, such as the Doris Marshall Institute, with 'white' racism against 'people of colour'. This conflation clearly removes from view many forms of racialised ethnic and religious antagonism and disadvantage. However, to recognise the diversity of racism and of white experience is not necessarily to deconstruct whiteness. The problem with the Doris Marshall Institute's interpretation is not merely its lack of sensitivity to the plurality of whitenesses but, more fundamentally, its faith in whiteness as a commonsense, obvious and discrete entity at the heart of racial history. Whiteness is thus employed as both the conceptual centre and the 'other' of anti-racism, the defining, normative, term of anti-racist praxis and theory. As this implies, to define whiteness, to acknowledge its contingent, slippery constructions, would radically destabilise orthodox anti-racism. It is towards studies that appear to promise such a transformation that I now turn.

Challenges and reaffirmations of anti-racist orthodoxy within 'white studies'

Despite the fact that discussion of racial whiteness is not new, since the 1980s the topic has begun to attract, if not quantitatively more attention, then certainly new forms of interest and from some new directions. This development – which, I should add, is largely confined to the USA – has enabled commentators to write about the formation of 'white studies'. One of the amusing characteristics of contributions to this field is that the term 'white studies' is usually assailed almost as soon as it is uttered. A nervousness about being seen to set up a specialism 'for whites', that 'indulges whites', hovers over the debate. 'My blood runs cold at the thought that talking about whiteness could lead to the development of something called white studies', asserts Dyer (1997, p.10), introducing his book *White*. '[W]e (arrogantly? narcissistically? greedily? responsibly?) believe that maybe this should be the last book on whiteness,' comment Fine *et al.* (1997, p.xii) in the opening pages of their edited collection *Off White*. I can think of no other area of study that has been, that could be, introduced in such a peculiar way. It is ironic that authors who invariably claim that their topic of enquiry is 'invisible' should seem to want to flee the scene of their own enquiries. This fear of the subject seems related to a failure to distinguish different ways of 'talking about whiteness'. The reification of whiteness, whether articulated through the celebration of white ethnicity or by other means, is, rightly, a cause for concern. But it is a concern that could only have come into view *because* of the existence of more critical approaches to the subject. The historicisation and spatialisation of whiteness – showing where it came from and how its forms vary across the world – are inherently antagonistic to any attempts to portray whiteness as a natural identity.

Before proceeding to differentiate and assess 'white studies', it may be useful to offer some explanations of why it has emerged in the closing decades of the twentieth century. It is important to note, first, that white studies emerged in the USA in the wake of the political and intellectual challenges offered by anti-racism and more radical versions of multi-culturalism. Such challenges have enabled a shift of emphasis within race equality work and debate away from non-white behaviour and attitudes and towards white racism. Thus white racism has been established as the key problematic of racial history. Although not lending itself directly to historical or geographical treatments of whiteness, this process has made white identity appear as a pressing issue, something that requires atten-tion. Indeed, in this sense, it may be said to have made whiteness more opaque, bringing it into focus as an object of concern and reflection.

A second stimulus for white studies has been the impact of deconstruc-tionist theories and themes within contemporary social science. Indeed, by exposing the reliance of 'what is privileged' upon 'what is marginal-ised', 'deconstructionism' has provided a specific analytical praxis and intellectual climate that has enabled and encouraged researchers to start interrogating the privileged centre of a number of different social arenas. Thus masculinity has been unpacked in gender studies, heterosexuality opened out in research on sexuality, and whiteness examined in ethnic and racial studies. As if to indicate the intellectual messiness and contra-dictory basis from which research projects can spring, the third of the influences on white studies that may be isolated relies on essentialist ideas that are theoretically incompatible with a deconstructionist approach. I am referring to that set of political agendas that constitutes, or is assumed to be constituting, white people as unable to experience 'real' racism, as inherently integrated within a racist system of colour privilege, and/or/thus as inevitably placed in an untenable position as commentators on racial oppression of non-white people (for discussion see Spivak, 1990; Asante, 1993). How ever exaggerated reports of this discourse may be (and for discussion of its inflation see Wright, 1994), it has lodged itself in the minds of many whites who consider themselves as anti-racist as a kind of omnipresent critique of their right to speak about 'others'. This sensitivity has, in part, been dealt with by white writers through the adoption of racially reflexive modes of address and analysis. The central device of the new introspective praxis is to monitor and acknowledge that one is speaking 'as a white person' (for a discussion of anti-racism and reflexivity see Bonnett, 1993b). To be able to speak in this manner implies, moreover, that one may authentically represent this group, that one has the right to speak about and, perhaps, even for, white people. In this sense, whites are enabled to take their place within the spectrum of identity politics. Thus, just as a similar essentialist rationality has encouraged some male gender researchers to identify masculinity as their appropriate subject matter, white scholars are finding that whiteness is a politically acceptable and alluring terrain on which to make 'their contribution'.

I shall now look at some recent examples of academic and activist work that seeks to raise explicitly the issue of whiteness within anti-racist debate. As we shall see, sensitivity to the importance of white racial identity in the struggle for race equality should not be confused with an anti-essentialist agenda. A divergence between the two tendencies is particularly apparent in the two forms of white studies I discuss first, the literature on anti-racism in white areas and anti-racist confessions of 'whiteness'.

Anti-racism in white areas

The anti-racist contention that, at least within white-dominated Western societies, racism is a 'white problem' clearly poses something of a challenge for a movement traditionally exclusively based in, and focused on, the most 'multi-racial' (i.e., least white) areas. In Britain in the mid-1980s, this challenge was taken up a by a wide variety of institutions and activists working within the so-called 'white highlands'.

The momentum of this activity was increased by mounting evidence that racist attitudes are as, or even more, deeply ingrained in white areas as they are in other places (Jay, 1992; Norcross, 1990; Troyna and Hatcher, 1992). In Britain, the release of the Swann Report (*Education for All*, Department of Education and Science, 1985) was also a significant stimulus. The report, which advocated the need for 'race' equality work in all parts of the United Kingdom, was used by a large number of Local Education Authorities (LEAs) to legitimise experiments in multicultural and anti-racist education. Indeed, in 1987, the Commission for Racial Equality (CRE, 1987) noted that 77 of Britain's 115 LEAs had produced a 'race' equality policy or were in the process of doing so.

Such publicly funded initiatives were significantly undermined by Conservative legislation limiting the roles and responsibilities of LEAs (see B. Taylor, 1990). However, the 'anti-racism in white areas movement' remains one of the most original, pragmatic and potentially influential movements within British anti-racism. Unfortunately, it may also be characterised by its refusal to interpret its central category, 'whiteness', as referring to anything other than a monolithic and stable racial entity.

Such assumptions may be introduced by turning to the work of one of the 'movement's' foremost proponents, Bill Taylor (1984; 1986; 1987; 1988). In a series of articles Taylor has argued for the need to adopt a 'gentle approach' when introducing multiculturalism and anti-racism in white areas. However, a less explicit argument also threads its way through Taylor's contributions. For Taylor repeatedly implies that not only does there exist a monolithic white identity but also that this entity has a 'monoculture'. Thus a triple conflation of region, culture and race occurs as Taylor comments on 'Britain's monocultural majorities' who live in what he terms 'non-contact zones'. Children and parents in such areas, he explains, 'are still cocooned in their essentially cultural homogeneity' (1986, p.180).

Revealingly, the characteristics of this 'white monoculture' have remained almost entirely undiscussed within the anti-racism in white areas debate. 'White culture' is referred to but never explained. Not unrelatedly, discussion of Britain's regional cultural diversity has been notable only for its absence. Taylor's work concerns itself primarily with Devon schools. Other contributions have focused on Cumbria (Brown *et al.*, 1990), Tyneside (Chivers, 1987) and Warwickshire (Sharma, 1987). Yet these 'multicultural' writings barely touch on the cultural distinctiveness of these places. The adoption of an all-encompassing notion of 'white culture' seems to suffocate such potentially destabilising thoughts.

Thus anti-racism has been developed in white areas as a way of encouraging white people to rethink their attitudes to non-white people. It has not sought to enable whites to understand themselves as racialised subjects. Nor has it attempted to explain why and how white people might have a stake in, or be able to engage with, anti-racism as a project that speaks to them about 'their own identities and histories'. The irony of the anti-racism in white areas debate is that it has been conceptually structured around the appraisal of representations of non-whites in the white imagination. Whiteness itself has been left untouched and fundamentally uninvolved.

Confessions of a white anti-racist

The recent flurry of interest in whiteness in both Britain and America has generally been viewed as a new and original phenomenon. However, there exist a number of pathways within this work, some of which are relatively well worn. The most significant of these more travelled routes is the literature and practice of white confession.

This paradigm, which seeks to enable and provoke white people to confront/realise/admit to their own whiteness, represents a reworking of the 'consciousness raising', or 'awareness training', forms of anti-racism that rose to prominence in the 1970s and early 1980s. These approaches were characterised by their interest in the way white people develop racial prejudices (for example, Wellman, 1977; Katz, 1978). More specifically, they tended to suggest that whites need to 'face up to' their own, and other white people's, racism in order to successfully expunge it from their psyche.

Over the past decade, a number of critiques of the individualistic, moralistic character of racism 'awareness training' (Sivanandan, 1985; Gurnah, 1984) have undermined its authority as a management and counselling resource. However, its confessional dynamic remains a potent force within anti-racism, including white studies. At its crudest, the confessional approach erases all questions relating to the contingent, slippery nature of whiteness. Instead a moral narrative is offered based on the presumed value of white 'self-disclosure' (see, for example, Chater, 1994; Camper, 1994). Thus, for example, in her article 'Biting the hand that feeds me: notes on privilege from a white anti-racist feminist', Nancy Chater (1994) attempts to expose her own, and other white feminists', whiteness to anti-racist critique. Drawing on that most reifying of reflexive devices, 'speaking as a ...', followed by a key-word mini-autobiography, Chater explains that 'as an anti-racist white feminist' she inevitably has

> to confront the ready potential of speaking or acting in ways that are based on or slide into arrogance, moralizing, self-congratulation, liberal politics, appropriation, careerism or rhetoric. (p.100)

Thus, whiteness is defined as referring to a racial group characterised by its moral failings, a community which is exhorted to be watchful of the reactionary tendencies apparently inherent in its anti-racist practice. Chater goes on to sketch prescriptively a number of ethical dilemmas faced by white anti-racist feminists. In particular, she suggests that such people need continually to monitor their seemingly innate capacity to silence non-whites and to acknowledge their own embeddedness within a racial elite. White feminists should also avoid assuming 'an edge of moral or intellectual superiority over and distance from other white people, especially those displaying a lack of politicised awareness of racism' (p.101). In other words, Chater is suggesting that whites should not – indeed cannot – escape being part of 'their racial group', or its attendant political conservatism.

I am not suggesting here that Chater's prescriptive strategies are necessarily wrong and that anti-racists should ignore or contradict her advice. What I wish to draw attention to is the process of category construction that structures her argument, and more specifically, the process whereby whiteness assumes a fixed and pivotal role as both a 'racial community' and a 'site of confession'. These locations establish whiteness as an arena, not of engagement with anti-racism but of self-generated altruistic interest for 'others' as well as for white people's own moral well-being. Indeed, it is tempting to argue that white confessional anti-racism establishes whiteness as the moral centre of anti-racist discourse. For whilst non-white anti-racists are cast as taking part in an instrumental politics of 'resistance' and 'self-preservation', 'white anti-racism' is continually elevated to a higher ethical terrain, removed from the realm of co-operation and participation to the more traditional (colonial, neo-colonial and anti-racist) role of paternalistic 'concern'.

The essentialising tendencies within Chater's text find echoes in other, more theoretically nuanced, work. For although, as we shall see, the development of historical geographies of whiteness has enabled a decisive move away from the reifying traditions outlined above, even some of its most adept adherents sometimes slip into moralising and confessional modes of analysis and address.

Thus, for example, Helen charles (1992) writes of the importance of 'Coming out as white'. Explaining that 'it is negatively exhausting teaching "white women" what it is like to be black', she notes that 'I feel it is now time for some verbal and textual "outness"' amongst white women (p.33). The implicit parallel charles is making with queer activists' 'outing' of closet homosexuals is an instructive one (see also Davy, 1997). It suggests that whiteness is being interpreted as a fixed disposition; a 'trait' that needs to be admitted to and exposed. And once uncovered, whiteness may presumably, be 'lived openly'. The association of whiteness with what is conceived to be a psycho-social proclivity (i.e., sexuality) undermines the anti-essentialist advances that have been made in recent white studies. More generally, I would conclude that the confessional approach forms a destabilising and unhelpful tendency when set against the more rigorously sociological trajectory being developed in other parts of the field.

The white working class and anti-racism

If one accepts that white identity needs to be understood as historically and geographically variable and, moreover, that its symbolic formation is, at least in part, bound up with the formation and reformation of capitalism, then one is likely to be critical of most existing forms of anti-racist practice and theory. It is not just that anti-racists tend to draw on (indeed rely on) a static, ahistorical conception of white identity. For just as damaging is their blindness to the changing class connotations of white identity, a myopia that leads to an interpretation of white identity as purely and simply an elite identity. The inadequacy of abstracting anti-racism from class critique has been pressed home by a number of recent commentators. Perhaps the most significant in a British context was the Burnage Report (commissioned and then suppressed by Manchester City Council) into the murder of Ahmed Ullah, a British Asian boy, by a white boy in the playground of Burnage High School on 17 September 1986. At the time of the murder the school had a number of anti-racist polices. Rightly or wrongly, it has been the nature and the supposed failure of these policies that provided the focus of the debate about Ahmed Ullah's death, becoming a touchstone within subsequent discussions on the future of anti-racism in Britain. The Burnage Report identified the typecasting of white working-class pupils and parents in the schools' anti-racism policy as central to the policy's failure. If, the report concluded, 'white students are all seen as "racist", simply by virtue of their whiteness, then anti-racism becomes merely a "moral" or "symbolic" exercise, (Macdonald *et al.*, 1989, p.347).

Since the assumption is that black students are the victims of the immoral behaviour of white students, white students almost inevitably become the 'baddies'. The operation of the anti-racist policies almost inevitably results in white students (and their parents) feeling 'attacked' and all being seen as 'racist', whether they are ferret-eyed fascists or committed anti-racists or simply children with a great store of human feeling and warmth who are ready to listen and learn and to explore their feelings towards one another ... Racism is placed in some kind of moral vacuum and is totally divorced from the more complex reality of human relations in the classroom.

This 'more complex reality' necessarily entails a class perspective. 'To deal with sex and race, but not class,' it continued (Macdonald *et al.*, 1989, p.33) 'distorts those issues ... This ostrich-like analysis of the complex of social relations leaves white working class males completely in the cold. They fit nowhere.' One of the authors of the report, Gus John (1991, p.85) went on to note 'the integral relationship between the struggle for social justice and the struggle for racial justice'. John (quoted by Young 1988, p.40) also reflected that 'the most successful anti-racist policy one can have is one that assumes that white parents have got to own racism as an issue within the school.'

Drawing on material derived from interviews with parents, the Burnage Report repeatedly highlighted white working-class resentment and frustration with the school's anti-racist policies. In particular, the report's authors noted a sense of grievance that anti-racist measures amounted to 'special treatment' and 'favoured treatment' of non-white pupils. White respondents also articulated a conviction that anti-racism produces racial resentment. One interviewee, Mrs Roscoe, noted that

I felt this enforced focus on multi-culturalism produced and produces prejudices. They are standing 5 year olds up in class pointing out differences for instance where people come from – and I feel this makes differences ... I know that some racism also comes from families themselves, but I do not feel that the multi-cultural approach helps because it makes the parents madder and more racist. For instance if kids go home to racist parents and talk about Pakistan or whatever, they would get a racist response. (Roscoe, cited by Macdonald *et al.*, 1989, p.325)

Concerns about the alienation of the white working class from anti-racism have been taken up by a variety of academic commentators (Cohen 1992; Gilroy 1987; Nelson 1990). Indeed, I am tempted to claim that, in the late 1980s and early 1990s, a new orthodoxy was in the making within British anti-racism, one that concluded that effective anti-racism cannot 'ignore', 'leave in the cold' or 'exclude' the white working class. Yet these formulations remained vague. They offered a vision of 'the white working class' engaging in the anti-racist project. But what does this mean? The Burnage Report suggested that educators should take time to 'explore' with white working-class youths their 'feelings or racist beliefs' (p.371). Other approaches have included offering this group a consciously non-threatening, 'white-inclusive', variety of anti-racism (for example, so-called 'softly-softly anti-racism', see Bonnett 1993a). Such strategies are certainly defensible as part of an overall and integrated vision of emancipatory education. However, we need to be careful that 'engaging' and 'listening to' do not become euphemisms for 'respecting' static, naturalised visions of an unchanging 'white culture'. Moreover, the plea that anti-racism is 'unfair to whites' requires particular attention. Gallagher's (1995) study of white racial attitudes amongst university students in the USA suggests that the notion that whites are the victims of racial equality measures can be employed as the foundation stone for the construction of a defensive white ethnicity. Reflecting, in part, the contradictory nature of the empirical evidence, other commentators on whiteness in the USA have suggested that whiteness, far from becoming a form of ethnicity, is being displaced by ethnicity (see Elba, 1990). However, it is surely significant that this process is also said to be driven by a sense of racial grievance. Jacobson has recently suggested that

> This disavowal of whiteness has become more pronounced in recent years, particularly around the question of affirmative action. The notion that Jews, Letts, Finns, Greeks, Italian, Slovaks, Poles or Russians are not *really* white has become suddenly appealing in a setting where whiteness has wrongly become associated with unfair *dis*advantage. (1998, p.280)

Anger that 'they' are taking advantage of 'us' (usually because some of 'us' are letting them) is a common discourse within ethnic and racial conflicts. Yet the roots of this anger can rarely be taken at face value, especially when the group claiming victim status is not in a socially or economically subordinate position in comparison to 'them'. This point can be elucidated further by reference to Roger Hewitt's (1996) interview-based study of white working-class youths' attitudes to racism and anti-racism in London. This work, published in the pamphlet *Routes of Racism* and designed to assist community and youth workers, has few pretensions to be offering in-depth analysis. However, it does provide a useful illustration of the problems of identifying and engaging 'white backlash'. Hewitt's white respondents evidence a clear disaffection with racial egalitarianism and, more specifically, with anti-racist practices, accusing them of privileging other races over whites and of unnecessarily racialising local society. Hewitt roots this attitude in equity practices themselves. Indeed, he goes on to imply that multiculturalism and anti-racism are responsible for white racial grievance. This linkage is certainly a provocative one but without a more comprehensive view of the anti-racist practices Hewitt's interviewees appear so bruised by it that it is difficult to offer any useful conclusion as to its validity. It is widely accepted by contemporary anti-racists in Britain that much anti-racist work in the 1980s was clumsy and essentialist. But a causal link between these failings and white resentment, or white racism, is not something that can simply be assumed, no matter how vociferously white complainants assert its validity.

Hewitt's linkage of race equity work and a white sense of racial unfairness has the unfortunate consequence of radically foreshortening the history of the latter. Indeed, Hewitt explicitly claims this grievance sensibility to be a recent phenomenon. 'In examining what *has* changed in the way white people both young and old, talk about race issues,' he writes, 'we find a radical new theme: the theme of 'unfairness'' (p.33). In Chapter 2, I offered an appraisal of the entry of the white working-class British into whiteness. This narrative provides an alternative perspective on white working-class claims to be being treated unfairly. It indicates, first, that working-class whiteness, although sharing with bourgeois whiteness the theme of superiority, developed in the mid–late twentieth century a characteristic emphasis on the ordinariness of whiteness, of whiteness being equated with the 'decent', the 'respectable', the 'everyday'. This identity was evoked in opposition to and defined in relation to both non-white 'immigrants' and middle-class cosmopolitanism. The power and appeal of this oppositional discourse were wrapped up with the 'gains' of working-class people in twentieth-century Britain. The so-called 'post-war settlement', the formation of the welfare state, were routinely articulated in the 1950s and 1960s in racialised terms, as 'our' welfare system that should be used to benefit 'us' and should not be exploited by 'them'. The 'decent, ordinary people' that constitute the axis of Enoch Powell's defences of white Englishness were an aggrieved people, a people who were being taken for a ride, a people whose identity needed to be mobilised (and constructed) in order to preserve their 'rights' (especially within the sphere of public provision) from exploitation by 'others'. Once we begin to consider the specificity of working-class whiteness in Britain, and its relationship to the wider political culture of mid- and late twentieth-century British class politics, it becomes apparant that white claims to 'unfairness' are far from new and are unlikely to be satisfactorily responded to by changing, or even junking, anti-racism. As long as British welfare capitalism, and more generally the British national project, is construed as 'ours', as a white project only extended on sufferance to 'them', then the discourse of 'unfairness' will continue to be central to white British identities.

Historical approaches, particularly ones that open up the contradictory nature of the politics of whiteness (for example, its role as axis of both working-class solidarity and working-class racism), may also help to answer the key dilemma for anti-racism that Hewitt poses. 'Avoiding reproducing the racializing process while at the same time tackling racism,' he notes (p.49) 'is crucial to any refocusing of anti-racism.' This is, indeed, a vitally important issue and I can only agree with Hewitt that it should be at the heart of any reconceptualisation of anti-racism. However, as I have sought to show, refocusing anti-racism should not arise out of, or lead towards, an overemphasis on the past failures of anti-racism. Anti-racism may have exacerbated white racial resentment in London, but it did not create it. Indeed, in the context of the nature of white working-class identity – in particular its ingrained oppositional and grievance-oriented tendencies – anti-racism was always likely to spark a reaction. Anti-racism was treading on very sensitive territory. Unfortunately, it is true to say that it often acted as if it was not, as if telling people that white racism was wrong would necessarily chime with some innate altruism. In fact, such an approach entirely misses the point of racism for many working-class whites; they did not use whiteness merely to feel superior, to claim they were extraordinary, but also, and perhaps more importantly, to feel part of, and defend, a notion of 'ordinary', 'decent' community. By opening up, by exploring, the contingent and contradictory nature of such identities anti-racism may usefully engage whiteness whilst refusing to treat white identities as static. The history related in Chapter 2 suggests that such an education must also enable students to explore the racialised nature of different forms of capitalism. This seemingly innocuous suggestion carries with it a range of complexities, some of which run directly counter to anti-racist orthodoxy. As I have suggested, within the British context, the most important of these complexities is the relationship between welfare capitalism and white identity. Thankfully, the relationship between welfarism, left-wing politics and racism has recently attracted the attention of a number of European commentators. More specifically, Broberg and Roll-Hansen (1997) have opened up a debate on the role of eugenics in Scandinavian welfare ideology (for discussion of this issue in Britain see Jacobs, 1985; Freedland, 1997; in China see Dikötter, 1995). This research will hopefully enable further historical and geographical discussion on the connections between socialist and racist ideologies. White identity and welfare intervention are inextricably interwoven 'gains' of 'European-heritage' working-class Britons. Working-class whiteness in Britain should be approached as a socioeconomic achievement of the working class that is mired in racism. In the twentieth century, the social formation that has enabled, and been enabled by, the symbolic constitution of white identity has combined social reformism (even, sometimes, class militancy) and racism. As this implies, a moment of social critique is contained within the history of white identity. Anti-racists should be engaged in the task of identifying and enabling this emancipatory dynamic, of harnessing it in the service of the transcendence of white identity, and its supersession with a politically defined identity. There is no room in such an approach for blaming or excluding the 'European-heritage' working class for 'being white'. Something more positive, and more daring, is required: something that encourages complex and contradictory responses to the question: 'What kind of achievement is whiteness?'

Becoming white: studies of white racialisation

This chapter has outlined a variety of ways anti-racists, whether ostensibly sensitive to white racial identities or not, have reified whiteness. In this final section I will introduce a countertendency. The writers and activists within this group offer an interpretation of whiteness characterised by two things:

1. an analysis of the social contingency of whiteness;
2. a critique of the category 'white', as currently constructed and connoted, as racist (but not necessarily a belief that all those people commonsensically assumed to be, or labelled, white are, *ipso facto*, racist).

The critical focus of this group is upon the racialisation process that produces whiteness. Their political problematic is how this process may be recognised without simultaneously being reproduced. Thus the existence of whiteness as a social fact is acknowledged, dissected and resisted. However, there exists considerable diversity within this school. Two broad tendencies may be discerned. The first attempts to subsume the analysis of whiteness within a class analysis of the racialisation process. The second stresses the plural constitution and multiple lived experiences of whiteness.

Theodore Allen (1994), David Roediger (1992; 1994), Noel Ignatiev (1995) and the contributors to the journal *Race Traitor* may be placed firmly within the former camp. Each traces whiteness as a project of US capitalism and labour organisations and each explicitly calls for its 'abolition'. These scholars and activists view white identity as the creation of racialised capitalism, an ideology that offers false rewards to one racialised faction of the working class at the expense of others. Thus it is argued that the task of anti-racists is not to encourage white people to confess to their 'own identity' but enable them to politically and historically contextualise, then to resist and abandon, whiteness. The editors of *Race Traitor* (1994) explain their project in the following terms:

> Two points define the position of *Race Traitor*: first, that the 'white race' is not a natural but a historical category; second, that what was historically constructed can be undone. (p.108)

Whiteness is presented here as an entirely oppressive identity. 'We will never have true democracy,' explains Rubio (1994, p.125) in the same journal, 'so long as we have a "white community"' (p.125).

However, the political conclusions of these historical studies derive, in the main, from a relatively limited reading of the synchronic social context of whiteness. For, although, whiteness is seen as, to use Allen's words, 'the overriding jetstream that has governed the flow of American history' (1994, p.22), it is analysed as if it were almost entirely a product of class, and particularly labour, relations. Thus, although a precise and useful account of the construction of whiteness emerges from these texts, it is not one that readily opens itself to dialogue with other histories or struggles.

Moreover, there exists within this body of work an unhelpful romanticisation of blackness. Indeed, *Race Traitor's* project is not merely to destroy whiteness but to enable whites to 'assimilate' blackness. Of course, blackness too is seen as a social construction. But it is construed as a construction that needs to be supported and reproduced. The editors argue that

> when whites reject their racial identity, they take a big step towards becoming human. But may that step not entail, for many, some engagement with blackness, perhaps even an identification as 'black'? Recent experience, in this country and elsewhere, would indicate that it does. (*Race Traitor*, 1994, p.115)

This formulation is clearly based upon a series of assumptions concerning the meaning of blackness. It implies that the romantic stereotype of the eternally resisting, victimised 'black community' is required to be further strengthened in order to create a suitable location for escapees from whiteness. Thus black people are condemned to reification as the price of white people's liberation from the racialisation process.

The somewhat clumsy political strategies of the white abolitionists are a disappointing conclusion to their work. However, the fact that the worries and constraints of political prescription are less apparent within other studies of white racialisation may help to explain why they appear so ready to 'complicate' the topic, resisting class reductionism in favour of an emphasis on the plurality of ways whiteness is conceived and enacted. A fine example of this latter trend is Matthew Jacobson's (1998) *Whiteness of a Different Color*. Jacobson's study focuses on the ethnic fissures within whiteness in US culture. In contrast to the attempt to root white identities in the history of labour struggle, Jacobson elucidates the diverse cultural, social and legal constitution of whiteness (on the legal construction of whiteness see also Lopez, 1996) and its continuing instability in the twentieth century. Ruth Frankenberg's *The Social Construction of Whiteness* also provides a number of insights into the slippery, incomplete and diverse nature of white racial identity. Frankenberg draws from her interviews with thirty white Californian women a complex portrait of the

> articulations of whiteness, seeking to specify how each is marked by the interlocking effects of geographical origin, generation, ethnicity, political orientation, gender and present-day geographical location. (p.18)

Thus, for example, in a chapter entitled 'Growing up White: The Social Geography of Race', Frankenberg explores her interviewees' childhood experiences of whiteness. For some, whiteness was always something explicit and physically and morally separate from non-whiteness. 'I grew up in a town,' explains one respondent, where 'everyone was aware of race all the time and the races were pretty much white people and Black people' (p.51). However, for another women, as Frankenberg notes, '"white" or "Anglo" merely described another ethnic group' (p.65). One interviewee, enculturated within a 'mixed' Mexican and white community, explains that she

> never looked at it like it was two separate cultures. I just kind of looked at it like, our family and our friends, they're Mexicans and Chicanos, and that was just a part of our life. (p.66)

Unfortunately, Frankenberg does not engage with the ambiguities of 'Hispanic' identity. Thus, for example, she ignores the – surely pertinent – fact that, as Henwood (1994, p.14) notes, in 'the 1990 Census half of all Hispanics reported themselves as white, a little under half as "other", and a few as black, native, or Asian'. However, despite this absence, Frankenberg's discussion of the multiple and shifting boundaries of whiteness is of immense value. It also provides a number of interesting points of contact with other studies of the hybrid nature of racial subjectivities.

Some instances of whites who 'cross over' into other racial identities were addressed in Chapter 4. The fact that these individuals were primitivists should provoke some circumspection as to the relationship between anti-racism and this form of 'racial treason'. Indeed, the best-known exposition on the latter topic, Norman Mailer's (1961; see also George, 1998) essay 'The white negro', which addressed young whites' emulation of African-American culture (more particularly, of jazz music and style), spoke of this sense of identification as a 'primitive passion'. 'Hip,' Mailer noted, 'is an affirmation of the barbarian.' However, it remains the case that in many circumstances the act of white racial identification with non-whites does represent a kind of social transgression (see Rubio, 1993; Rogin, 1996). A longer view of the formation and politics of the 'wigger' (white nigger) is offered by Roediger (forthcoming), who charts the connections between black-identifying and/or outcast elements (for example, so-called 'white trash', see also Wray and Newitz, 1997) of the white working class from the late nineteenth century. In Britain, the most sustained study of white youth identification with blackness is Simon Jones's (1989) ethnographic study of white Rastafarians in Birmingham, a 'white community' that self-consciously splices its own whiteness with styles and ideologies associated with Rastafarianism. This escape, as Jones notes, draws on a correlation of whiteness with boredom and passivity and of blackness with rebellion and the exotic. It is an 'escape', then, based on certain familiar clichés of whiteness and blackness. However, despite this reliance, the process of becoming and socially interacting as a white Rastafarian inevitably opens up the fluidity of racial identity, creating incomplete, impermanent and explicitly constructed moments of appropriation and cultural 'play'. As this implies, the notion that the only escape from whiteness is into another racial monolith, blackness, exists in a tense relationship with the postmodern emphasis on hybridity. Reflecting the influence of the latter paradigm, the creative appropriation and intermixing of ethnic identities have increasingly become a central focus of commentators on contemporary youth cultures (for example, Hebdige, 1979; Ross and Rose, 1994; Nayak, 1997; 1999). One such moment of hybridity is addressed by Jeater (1992) in her account of the 'multi-racial' anti-racist politics of late-1970s inner London. At that time, recalls Jeater,

> we all began to celebrate the complexities and interdependencies of our cultural heritages. White people like myself ... who grew up listening to reggae music and who perhaps took part in the urban uprisings against the state, were as much a part of this project as everyone else ... the cosmopolitanism and the dynamic interactions of cultural traditions created a real sense that the world was there to be forged in new way. (pp.118–19)

Such moments of crisis and youth revolt provide perhaps the clearest indications of the possibility of the deconstruction of racial categories, of creative hybridity. However, as the work already cited in this section implies, the 'confusion' and intermixing of racial signs and boundaries are not restricted to moments of youthful transgression. Disruptive and mutant forms of 'white' identity have a long and varied lineage. As Sakamoto (1996) and Young (1995) have shown, hybridity is, in itself, neither transgressive nor necessarily non-racist. However, within the context of an identity like whiteness, which is so often treated as a static, unchanging monolith, the ability to identify and trace moments of inter-mixture and creative adaptation provides a necessary resource, a way of opening up whiteness and imagining its transformation or supersession.

The notion that whiteness is a dominant identity and, hence, an invisible identity is found throughout the 'white studies' literature. It is also widely accepted that those groups who are the victims of white racism are the ones most likely to be able to 'see' whiteness. Reflecting the dominance of the USA in this debate, the vantage point which has most frequently been cited as having a 'good view' on whiteness is that of the African-American. Roediger's (1998a) edited collection *Black on White: Black Writers on What it Means to be White* offers a wealth of African-American commentary on whiteness. The earliest piece in Roediger's collection – David Walker's (1998, p.54) essay 'Whites as heathens and Christians', published in 1830 – offers a scathing attack on whites as 'an unjust, jealous, unmerciful, avaricious and blood-thirsty set of beings'. Roediger's compilation clearly supports the point made by bell hooks (1998) in her influential essay 'Representations of whiteness in the black imagination', that whiteness has long been monitored by African-Americans. Indeed, it is a contention that was articulated with some force by the most eminent US intellectual of the twentieth century, W.E.B. Du Bois. In a chapter in *Darkwater* (1920) titled 'The souls of white folk', Du Bois notes of whites:

> Of them I am singularly clairvoyant. I see in and through them. I view them from unusual points of vantage. Not as a foreigner do I come, for I am native, not foreign, bone of their thought and flesh of their language. Mine is not the knowledge of the traveller or the colonial composite of dear memories, words and wonder. Nor yet is my knowledge that which servants have of masters, or mass of class, or capitalist of artisan. Rather I see these souls undressed and from back and side. I see the working of their entrails. I know their thoughts and they know that I know. This knowledge makes them now embarrassed, now furious. They deny my right to live and be and call me misbirth! (1998, p.184; first published 1920)

Another African-American writer, James Baldwin (1983, p.167), noted that this knowledge of whiteness was the product of white racism: 'it is one of the ironies of black–white relations that, by means of what the white man imagines the black man to be, the black man is enabled to know who the white man is.' Crispin Sartwell's (1998) recent attempt to write about his own white identity by reference to black autobiography offers further expression of this conviction:

> The oppressor may appeal to many noble principles, may himself take himself to believe them, but the oppressed know the truth about the oppressor's deepest self, which is constantly expressed in oppression ... Black folks wouldn't countenance the sort of stuff we white folks believe about ourselves, but we white folks can't allow ourselves to know what black folks know about us. That is why our only hope for ceasing to be oppressors is to listen to what black folks are saying about us. (p.83)

Sartwell's use of the male pronoun for the 'oppressor' is intentional, and suggestive of the existence of another social location that might be imagined to offer insights into whiteness, that of women. However, what is most striking about this passage are its generalisations of racial sentiment. What 'black folks' and 'white folks' think is, apparently, obvious, commonsense knowledge. Thus, although Sartwell's suggestion that whites need to listen to the voices of the victims of white racism is to be applauded, he seems to have already made up his mind as to what such voices would reveal. For Sartwell, 'our [i.e., 13white] own fantastic psychopathology' boils down to the fact that 'we were destroyers: not Christians, not democrats, but destroyers' (p.83). The Manichean conflict between good and evil that Sartwell portrays suggests that any privileging of non-white perspectives in anti-racism needs to proceed with care. The direct experience of white racial oppression clearly does provide a valuable vantage point from which to see the workings of white racism, especially when, as with Du Bois, it is allied with a sensitivity to the possibility and process of moving between different realms of racial experience. However, this fact should not be confused with the notion that merely by virtue of being excluded and marginalised one is enabled to comprehend and to make sense of one's oppression. It is pertinent to note here that the philosophical abstractions that dominate Sartwell's study and the wider academic debate – whether dialectical, existential or deconstructionist – have the unfortunate effect of removing from view the social reproduction of ideology. For, as innumerable commentators on class consciousness have suggested, one of the effects of oppression is that the oppressed are kept in ignorance and, moreover, that they are susceptible to obfuscatory and divisive beliefs, such as religion and racial essentialism. These ideological beliefs mean that, although exploited people may see and feel their oppression, they do not necessarily understand its causes or the possibility that it can be overthrown. There is, moreover, another way that a philosophical perspective – more especially a *Western* philosophical perspective – provides a restrictive and distorting interpretative structure through which to address sociological problems. More specifically, the dualism of the Western philosophical tradition (master and slave, self and other, centre and periphery) undermines more fluid and multifaceted analysis of 'race relations'. Although this philosophical structure has helped sustain the already highly binary nature of the race debate in the USA, in other societies (and increasingly in the USA itself), the notion that only two positions are available within racial discourse appears unhelpful. In the survey of non-white responses to whiteness presented in Chapter 4 a variety of stances were noted, from accommodation through to resistance. It is also relevant to recall that although the development of ideologies of occidentalism in Latin America and East Asia may be taken to have enabled whiteness and Westernness to be 'seen', this process did not necessarily entail the acquisition of some superior or more informed knowledge of those identities. However, I would also suggest that once the multiple ways whiteness is interpreted around the world are engaged, the notion that looking at whiteness from non-white perspectives offers a genuine challenge to white identity begins to make much more sense. Rather than confirming whiteness by placing it in a closed binary circuit of political positions (in which the only escape from whiteness is into a racist fantasy of blackness), such an international and historical frame of reference disperses whiteness, shows how it can be mutated and how lives can be lived without it.

Conclusions: engaging whiteness

This chapter has focused on one of the most intractable, and I believe counterproductive, of anti-racism's traditional monoliths. It has argued that anti-racism has objectified whiteness and that this process has been perpetuated within the 'white areas' debate and confessional approaches to anti-racism. However, as we have seen, the past few years have also witnessed signs of a new willingness to look at the historical and geographical contingency of whiteness. This latter body of work enables a reconceptualisation of whiteness as a diverse and mutable social construction. Clearly, this trajectory also implies a new level of sophistication in both the recognition of, and resistance to, whitenesses.

I wish to conclude this chapter by engaging with the debate around the meaning, as well as the political and social implications, of an anti-racist, anti-essentialist perspective on white identity. Within the academic field of racial studies, most of the English-language debate about the merits and demerits of anti-essentialism has focused on black identity. More specifically, controversy has been aroused by the work of poststructuralist African-American writers (for example, Gates, 1986; 1988; Baker, 1986; Appiah, 1985; see also Fuss, 1989; Abel, 1993), who have seemed to undermine the meaning and political coherence of blackness. To deconstruct blackness, it has been argued, is politically naive. 'It is insidious,' notes Joyce (1987a, p.341; see also Joyce 1987b), 'for the Black literary critic to adopt any kind of strategy that diminishes or ... negates his blackness.' As Fuss (1989, p.77) points out, critics such as Joyce charge that 'to deconstruct "race" is to abdicate, negate, or destroy black identity.' Identifying poststructuralism with white critics, Fuss argues that

> In American culture, 'race' has been far more an acknowledged component of black identity than white; for good or bad, whites have always seen 'race' as a minority attribute, and blacks have courageously and persistently agitated on behalf of 'the race'. It is easy enough for white poststructuralist critics to place under erasure something they *think* they never had to begin with. (p.93)

In the eyes of some critics, the gulf between the deconstructive and essentialist positions has been bridged by the development of 'a strategic use of positivist essentialism in a scrupulously visible political interest' (Spivak, quoted by Fuss, 1989, p.31; see also Baker, 1986). Such a position enables minority groups to 'preserve' identities that facilitate struggle, resistance and solidarity whilst maintaining a critique of reified notions of 'race'. Asked to expand upon the implications of strategic essentialism, the term's progenitor, Spivak (1990, p.93), comments, 'the only way to work with collective agency is to teach a persistent critique of collective agency at the same time ... It is the persistent critique of what one cannot not want.'

Spivak's allusion to 'what one cannot not want' reinforces the impression that this is a debate formulated entirely around the perceived interests of oppressed groups (more specifically, African-Americans). As this implies, the question of when we should stick to blackness and when we should critique it cannot simply be translated onto questions of whiteness. Given the exclusionary and normative nature of its development, any form of essentialist 'sticking to' whiteness is not a viable anti-racist position. As we have seen, whiteness has developed, over the past two hundred years, into a taken-for-granted experience structured upon a varying set of supremacist assumptions (sometimes cultural, sometimes biological, sometimes moral, sometimes all three). Non-white identities, by contrast, have been denied the privileges of normativity and are marked within the West as marginal and inferior.

Unfortunately, those seeking to develop arguments for, or counter-arguments against, a politically engaged anti-essentialism have rarely considered the implications of these positions for white identity (though see Abel, 1993). Thus some of the most important questions for an anti-foundationalist anti-racism have remained undiscussed. Perhaps the most pertinent of these is how whiteness can be made visible, presented for critical inspection, whilst at the same time exposed as a myth, a racist construction that needs to be, if not abolished, permanently caged between inverted commas. In other words, we need to ask how the enormous power of whiteness (through white institutions, power dynamics, individuals, etc.), can be acknowledged and confronted at the same time as its essentialist pretensions are denied.

It is important to note that the central tension at work within these questions is not between essentialism and anti-essentialism. Acknowledging the social power, the social existence of whiteness is not the same as claiming (however ironically or self-consciously) that it is a fixed or natural category. As this implies, the position that bridges the tension outlined in the questions I ask above may more usefully be termed 'strategic deconstruction' than 'strategic essentialism'. The problematic of strategic deconstruction is not when and how to 'stick to', 'preserve' or 'save' whiteness, but when and how should whiteness be opened up, torn open, made slippery and when and how should it be revealed and confronted?

One possible route out of this dilemma is to view whiteness as a political category. 'Black' has long been used to incorporate and cohere a trans-racial community of resistance. The 1805 Constitution of Haiti declared all Haitians were black, no matter what their skin colour was. 'Now when I say black,' asserted one of the founders of Black Power, Malcolm X (1987, p.12) in March 1964, 'I mean non-white. Black, brown, red or yellow.' Ever since James Baldwin's provocative assertion that 'As long as you think you are white, there's no hope for you' (quoted by Roediger, 1994), a political reading of whiteness has remained a minor theme within the most incisive anti-racist work. In the mid-1980s, Clark and

Subhan (undated, p.33) suggested, but did not develop, the notion that 'Both in global terms and in the British context ... White as a political term is a term for the oppressor.' Political whiteness may most usefully be viewed as an intellectual resource rather than a universal solution. In certain circumstances it may provide an appropriate way of approaching white identity that is able to remain both theoretically and practically adroit. However, in other contexts, political whiteness may appear too resolutely negative, offering white people nothing but a set of indulgent guilt complexes and erasing the multiple and fragmented ethnicities that overlap with whiteness. As many commentators have observed (Macdonald *et al.*, 1989; Cohen, 1992), one of the most important tasks of contemporary anti-racism is to engage white people, to bring them 'inside' the anti-racist project. This implies that the notion of political whiteness should be set within a wider and more sophisticated anti-racist project that enables the historical, international and personal experiences of whiteness to be explored in the context of a changing global economy. Such a process could provide white people with a stake in anti-racism, as a project that talked to and about them, whilst weakening the common-sense, normative nature of white identity. Whiteness has traditionally been the invisible centre of the 'race' equality debate. It is now time to draw it into an explicit *engagement* with the anti-racist project.

A concluding remark

It is a characteristic conceit of the modern era that each generation, indeed each decade, is heralded as transformatory. The notion that 'we live in extraordinary times' is recycled again and again, often attended by a flotilla of similarly alluring slogans, clichés that suggest that today is a time of 'crisis', of 'rapid change', of 'bewilderment'. I once considered 'whiteness' in such terms. There was, I felt, a 'crisis' in white identities, something unique, something entirely new. I was, in part, influenced in this opinion by nothing more profound than the fact that books were now being published on the subject and the media in the USA were announcing 'white studies' as a hot new topic. In this brave and novel moment, white identity was being 'questioned' as never before, held up for 'critical scrutiny' for the very first time.

Of course, once the study of white identities is viewed as a fashion it must be accorded its fifteen minutes of fame and then, like all fashions, fade. After all, if the questions, the scrutiny, appeared so suddenly they can, presumably, disappear just as quickly. What inevitably escapes such a fashion-led agenda is the way that whiteness has been observed, been adopted and adapted, undergone 'crisis' and 'rapid change' in numerous ways in diverse parts of the world for centuries. It has not suddenly come into being as a topic of enquiry and controversy in the 1990s. Moreover, if I have been right in arguing that white identity and modernity have been mutually constitutive, and that the former has acted to naturalise the latter, then it seems unlikely that whiteness will disappear any time soon. The end of race, the end of racism and the demise of whiteness have been announced on several occasions over the past decade. Common reasons offered for these fatalities are the intellectual bankruptcy of scientific racism and the increasing salience of other kinds of social cleavage. Perhaps a useful first response to such claims is to go and take a look. Does whiteness no longer matter to people in Latin America; is the category *'gaijin'* no longer synonymous with white Westerners in Japan; are the five terms 'European', 'white', 'Westerner', 'developed', and 'advanced' no longer associated? There is more than enough empirical evidence to suggest that whiteness remains a core component of contemporary political and social identities around the world. Such empirical evidence might also, however, cause us to question the terms of a debate that postulates the essence of race in bad biology. For to say merely that whiteness 'remains' important, to assert simply that it is 'still with us', is to be drawn into placing whiteness as an example of social continuity, a static practice and ideology to be contrasted with practices and ideologies that reflect social transformation. I have been arguing in this book that if one understands the development of white identities in their historical and international context then one can also begin to appreciate that they have been forged in a dynamic relationship with modernity. This relationship implies that white identity does not have an unchanging, narrowly defined core theory that can be tracked down (for example, biological racism) and pronounced dead or alive. Rather it has relied on much broader processes of naturalisation and reification, processes that can be theoretically rearticulated for each new generation. In Chapter 4, I suggested that, far from being a decaying legacy from the past, the social ideal of whiteness in Latin America was being reshaped within the context of neo-liberalism. Whiteness has increasingly become associated with the ideal of consumerism, allied to individualist and competitive ideologies of 'choice' and 'personal freedom'. This form of whiteness exists alongside other, more traditional, visions of white identity. However, these recent incarnations represent a clear reflection of the fact that white identities are not necessarily rooted in the era of racial science, and that they will not automatically be surpassed simply because the racial theories of the nineteenth century are no longer given credence. If we accept that modernity has been naturalised through ideologies of whiteness, it should come as no surprise that the two continue to be intertwined, albeit in new and changing ways. This relationship throws up a considerable challenge for anti-racists. The task is not simply one of opposing whiteness. Rather it is one of deracialising and, more fundamentally still, denaturalising modernity.

I was born white; so were my parents and grandparents. For all of us the world was dominated by Europeans. It was Europeans who were shaping the planet and its peoples. And it was white skin that symbolised who was European. The ultimate marker of power and modernity was a symbol of the natural, a sign beyond dispute. Thus the ideology of racial whiteness transformed political and economic processes into 'facts of life', things that were above question and beyond challenge. It is my hope that by engaging and exploring the history and geography of whiteness we can help overturn the meaning of modernity.

References

Abel, E. (1993) 'Black writing, white reading: race and the politics of feminist interpretation', *Critical Inquiry*, **20**, 3, pp.470–98

Acosta, J. (1604) *The Natural and Morall Historie of the East and West Indies*, London,Val. Sims for Edward Blount and William Apsley

Adair, M. (1992) 'Will the real men's movement please stand up?' in L. Hagan (ed.) *Women Respond to the Men's Movement*, San Francisco, Pandora

Allen, G. (1979) 'Are we Englishmen?', in M. Biddiss (ed.) *Images of Race*, Leicester, Leicester University Press

Allen, T. (1994) *The Invention of the White Race*, Volume One: *Racial Oppression and Social Control*, London, Verso

Ampiah, K. (1995) 'Japan at the Bandung Conference: The cat goes to the mice's convention', *Japan Forum*, **7**, 1, pp.15–24

Anon. (1992) 'Sacred event or "goofy circus"', in C. Harding (ed.) *Wingspan: Inside the Men's Movement*, New York, St Martin's Press

Appiah, A. (1985) 'The uncompleted argument: DuBois and the illusion of race', *Critical Inquiry*, **12**, 1, pp.21–37

Asante, M. (1993) 'Where is the white professor located?' *Perspectives*, **31**, 6, p.19

Azzam, A. (1976) 'The Arab League and world unity', in S. Haim (ed.) *Arab Nationalism: An Anthology*, Berkeley, University of California Press

Baker, H., Jr (1986) 'Caliban's triple play', *Critical Inquiry*, **13**, 1, pp.182–96

Baldwin, J. (1983) *Notes of a Native Son*, Boston, Beacon Press

Banton, M. (1987) *The Idea of Race*, London, Tavistock

Banton, M. (1987) *Racial Theories*, Cambridge, Cambridge University Press

Barcata, L. (1968) *China in the Throes of the Cultural Revolution*, New York, Hart Publishing Co.

Bastide, R. (1968) 'Color, race, and Christianity', in J. Franklin (ed.) *Color and Race*, Boston, Houghton Mifflin

Bauman, Z. (1997) 'No way back to bliss', *Times Literary Supplement*, 24 January

BBC (1998) *Kali's Smile*, broadcast on Radio Four, 6 May

Beddoe, J. (1885) *The Races of Britain: A Contribution to the Physical Anthropology of Western Europe*, Bristol, J. W. Arrowsmith

Bédoucha, G. (1982) 'Citer un blasphème n'est pas blasphémer': du racisme en pays d'Islam', *Cahiers d'Études Africaines*, **22**, 3/4, pp.533–7

Beezer, A. (1993) 'Dark and fair women: race and gender in Haggard's African adventure narratives', *Cultural Studies from Birmingham*, **2**, pp.7–31

Bell, D. (1999) 'What does Confucius add to human rights?' *Times Literary Supplement*, 1 January

Benedict, R. (1940) *Race: Science and Politics*, New York, Modern Age Books

Bernstein, I., et al. (1981) 'Chinese and white concepts of attractiveness', *Bulletin of the Psychonomic Society*, **18**, 2, p.59

Bhabha, H. (1990) 'Interview with Homi Bhabha: The Third Space', in J. Rutherford (ed.) *Identity: Community, Culture and Difference*, London, Lawrence & Wishart

Blacker, C. (1969) *The Japanese Enlightenment: A Study of the Writings of Fukuzawa Yukichi*, Cambridge, Cambridge University Press

Blanch, M. (1976) 'Imperialism, nationalism and organized youth', in J. Clarke, C. Critcher and R. Johnson (eds) *Working-Class Culture: Studies in History and Theory*, London, Hutchinson

Bland, L. (1982) '"Guardians of the race", or "Vampires upon the nation's health"?: female sexuality and its regulation in early modern twentieth-century Britain', in E. Whitelegg (ed.) *Changing Experience of Women*, London, Martin Robertson

Bliss, S. (1992) 'What happens at a mythopoetic men's weekend?' in C. Harding (ed.) *Wingspan: Inside the Men's Movement*, New York, St Martin's Press

Blumenbach, J. (1997) 'The degeneration of races', in E. Eze (ed.) *Race and the Enlightenment: A Reader*, Oxford, Blackwell

Bly, R. (1990) *Iron John: A Book About Men*, Reading, Mass., Addison-Wesley

Bly, R. (1991) 'The need for male initiation', in K. Thompson (ed.) *To Be A Man: In Search of the Deep Masculine*, New York, Jeremy P. Tarcher/Perigee Books

Bly, R. (1996) *The Sibling Society*, London, Hamish Hamilton

Boer, P. (1995) 'Essay 1: Europe to 1914: The making of an idea', in K. Wilson and J. Dussen (eds) *The History of the Idea of Europe*, London, Routledge

Booth, W. (1976) 'Why darkest England?' in P. Keating (ed.) *Into Unknown England, 1866–1913: Selections from the Social Explorers*, Manchester, Manchester University Press

Bonnett, A. (1993a) *Radicalism, Anti-racism and Representation*, London, Routledge

Bonnett, A. (1993b) 'Contours of crisis: anti-racism and reflexivity', in P. Jackson and J. Penrose (eds) *Constructions of Race, Place and Nation*, London, UCL Press

Bonnett, A. (1994) 'The new primitives', *Variant*, **16**, pp.54–7

Bonnett, A. (1999) *Anti-racism*, London, Routledge

Brantlinger, P. (1986) 'Victorians and Africans: the genealogy of the myth of the dark continent', in H. Gates (ed.) *'Race,' Writing and Difference* Chicago, University of Chicago Press

Bray, R. (1907) *The Town Child*, London, Fisher Unwin

Broberg, G. and **Roll-Hansen, N.** (eds) (1997) *Eugenics and the Welfare State: Sterilization Policy in Denmark, Sweden, Norway, and Finland*, East Lansing, Michigan State University

Brod, H. (1992) 'The mythopoetic men's movement: a political critique', in C. Harding (ed.) *Wingspan: Inside the Men's Movement*, New York, St Martin's Press

Brooke, J. (1990) 'Brazil's idol is a blonde, and some ask "why?"', *New York Times*, 31 July

Brown, C., et al. (1990) *Spanner in the Works: Education for Racial Equality and Social Justice in White Schools*, Stoke-on-Trent, Trentham Books

Brown, L. (1968) 'Color in North Africa', in J. Franklin (ed.) *Color and Race*, Boston, Houghton Mifflin

Buffon, G. L., Comte de (1997) 'The geographical and cultural distribution of mankind', in E. Eze (ed.) *Race and the Enlightenment: A Reader*, Oxford, Blackwell

Burroughs, E. (1963) *The Return of Tarzan*, New York, Ballantine

Campbell, B. (1995) 'Little Beirut', *The Guardian*, 1 July

Camper, C. (1994) 'To White Feminists', *Canadian Woman Studies*, **14**, 2, p.40

Carrell, J. (1994a) About *Ravenflight* and the people behind it', *RavenFlight: A Journal of Men's Arts and Mysteries*, **1**, p.3

Carrell, J. (1994b) Of green men, goddesses, and the (many) men's movements: conversations in a tavern', *RavenFlight: A Journal of Men's Arts and Mysteries*, **1**, pp.36–9

CCCS (1981) *Unpopular Education: Schooling and Social Democracy in England since 1944*, London, Hutchinson

Chamberlin, J. (1994) 'Of green men, goddesses, and the (many) men's movements: conversations in a tavern', *Ravenflight*, **1**, pp.36–9

Champley, H. (1936) *White Women, Coloured Men*, London, John Long

Chapoval, T. (1992) 'Brazil's blond bombshell explodes worldwide', *San Diego Union*, 31 January

charles, H. (1992) 'Whiteness – the relevance of politically colouring the "non"', in A. Phoenix and J. Stacey (eds) *Working Out: New Directions for Women's Studies*, London, Falmer Press

Chater, N. (1994) 'Biting the hand that feeds me: notes on privilege from a white anti-racist feminist', *Canadian Woman Studies*, **14**, 2, pp.100–4

Chevalier, J. and **Gheerbrant, A.** (1996) *The Penguin Dictionary of Symbols*, London, Penguin Books

Chivers, T. (ed.) (1987) *Race and Culture in Education: Issues Arising from the Swann Committee Report*, London: NFER-Nelson

Clark, G. and **Subhan, N.** (undated) 'Some definitions', in K. Ebbutt and B. Pearce (eds) *Racism in Schools: Contributions to a Discussion*, London, Communist Party of Great Britain

Clifford, J. (1988) *The Predicament of Culture: Twentieth-century Ethnography, Literature and Art*, Cambridge, Mass., Harvard University Press

Cohen, P. (1988) 'The perversions of inheritance: studies in the making of multi-racist Britain', in P. Cohen and H. Bains (eds) *Multi-racist Britain*, Basingstoke, Macmillan

Cohen, P. (1992) '"It's racism what dunnit": hidden narratives in theories of racism', in J. Donald and A. Rattansi (eds) *'Race', Culture and Difference*, London, Sage

Cohen, P. (1997) 'Laboring under whiteness', in R. Frankenberg (ed.) (1997) *Displacing Whiteness*, Durham, NC, Duke University Press

Cohen, S. (1985) 'Anti-semitism, immigration controls and the welfare state', *Critical Social Policy*, **5**, 1, pp.73–92

Cohen, S. and **Taylor, L.** (1992) *Escape Attempts: The Theory and Practice of Resistance to Everyday Life*, second edition, London, Routledge

Collier, R. (1994) '"Back to the woods and on to the streets": change and continuity in ideas of masculine "crisis" and "renewal"', paper presented to the Critical Legal Conference, University of Warwick, September

Colls, R. (1986) 'Englishness and the political culture', in R. Colls and P. Dodd (eds) *Englishness: Politics and Culture 1880–1920*, London, Croom Helm

Colls, R. and **Dodd, P.** (eds) (1986) *Englishness: Politics and Culture 1880–1920*, London, Croom Helm

CRE (Commission for Racial Equality) (1987) *Learning in Terror*, London: CRE

Connell, R (1995) *Masculinities*, Cambridge, Polity Press

Coomaraswamy, A. (1956) *Medieval Sinhalese Art*, New York, Pantheon

Cox, O. (1948) *Caste, Class and Race: A Study in Social Dynamics*, New York, Monthly Review Press

Crampton, R. (1994) 'Dances with wolves', *The Times Magazine*, 6 August

Creighton, M. (1995) 'Imaging the other in Japanese advertising campaigns', in J. Carrier (ed.) *Occidentalism: Images of the West*, Oxford, Oxford University Press

Creighton, M. (1997) 'Soto Others and uchi Others: imaging racial diversity, imaging homogenous Japan', in M. Weiner (ed.) *Japan's Minorities: The Illusion of Homogeneity*, Routledge, London

Crosby, E. (1901) 'The military idea of manliness', *The Independent*, **53**, pp.873–5

Crossley, P. (1990) 'Thinking about ethnicity in early modern China', *Late Imperial China*, **11**, 1, pp.1–34

Crowhurst, A. (1997) 'Empire theatres and the empire: the popular geographical imagination in the age of empire', *Environment and Planning D: Society and Space*, **15**, pp.155–73

Curtis, L. (1997) *Apes and Angels: The Irishman in Victorian Caricature*, revised edition, Washingon, Smithsonian Institution Press

Dale, J. (1988) *The Myth of Japanese Uniqueness*, Routledge, London

Daly, T. (1992a) 'At a men's dance', *Thunder Stick: The Journal of Vancouver M.E.N.*, **2**, 2, p.5

Daly, T. (1992b) 'Initiation', *Thunder Stick: The Journal of Vancouver M.E.N.*, **2**, 3, pp.14–15, 20–21

Darwin, C. (1901) *The Descent of Man and Selection in Relation to Sex*, new edition, London, John Murray

Davis, F. (1991) *Who is Black? One Nation's Definition*, University Park, Pennsylvania State University Press

Davy, K. (1997) 'Outing whiteness: a feminist/lesbian project', in M. Hill (ed.) *Whiteness: A Critical Reader*, New York, New York University Press

Department of Education and Science (1985) *Education For All: Committee of Inquiry into the Education of Children from Ethnic Minority Groups*, London, HMSO

de Friedemann, N. and **Arocha, J.** (1995) 'Columbia', in Minority Rights Group (eds) *No Longer Invisible: Afro-Latin Americans Today,* London, Minority Rights Publications

Dickson, L. (ed.) (1938) *The Green Leaf. A Tribute to Grey Owl,* Lovat Dickson, London

Dickson, L. (1939) *Half-Breed: The Story of Grey Owl,* Peter Davies, London

Dickson, L. (1976) *Wilderness Man: The Strange Story of Grey Owl,* Sphere Books, London

Dikötter, F. (1990) 'Group definition and the idea of "race" in modern China (1793–1949)', *Ethnic and Racial Studies,* **13**, 3, pp.420–31

Dikötter, F. (1992) *The Discourse of Race in Modern China,* Stanford, Stanford University Press

Dikötter, F. (1994) 'Racial identities in China: context and meaning', *The China Quarterly,* **138**, pp.404–12

Dikötter, F. (1995) *Sex, Culture and Modernity in China,* London, Hurst

Dikötter, F. (ed.) (1997) *The Construction of Racial Identities in China and Japan,* London, Hurst

Dirlik, A. (1993) 'Review article: the discourse of race in modern China', *China Information,* **7**, 4, pp.68–71

Doris Marshall Institute (1994) 'Maintaining the tensions of anti-racist education', *Orbit,* **25**, 2, pp.20–21

Du Bois, W. (1920) *Darkwater,* New York, Harcourt, Brace and Howe

Du Bois, W. (1989) *The Souls of Black Folk,* New York, Penguin

Du Bois, W. (1998) 'The souls of white folk', in D. Roediger (ed.) *Black on White: Black Writers on What it Means to be White,* Schocken Books, New York

Dyer, R. (1997) *White,* London, Routledge

Eckersley, R. (1989) 'Green politics and the new class: selfishness or virtue?' *Political Studies,* **37**, pp.205–23

Elba, R. (1990) *Ethnic Identity: The Transformation of White America,* New Haven, Conn., Yale

Ellis, A. (1976) *Sex and the Liberated Man,* Secaucus, Lyle Stuart

Epperson, T. (1997) 'Whiteness in early Virginia', *Race Traitor,* **7**, pp.9–20

Faber. R. (1971) *Proper Stations: Class in Victorian Fiction,* London, Faber and Faber

Fanon, F. (1967) *The Wretched of the Earth,* Harmondsworth, Penguin Books

Fanon, F. (1986) *Black Skin, White Masks,* London, Pluto Press

Featherstone, S. (1998) 'The Blackface Atlantic: interpreting British minstrelsy' *Journal of Victorian Culture*, **3**, 2, pp.234–51

Fee, D. (1992) 'Masculinities, identity and the politics of essentialism: a social constructionist critique of the men's movement', *Feminism and Psychology*, **2**, 2, pp.171–6

Fine, M., Weis, L., Powell, L. and **Wong, L.** (eds) (1997) *Off White: Readings on Race, Power, and Society*, Routledge, New York

Finger, W. (1992) 'Finding the door into the forest', in C. Harding (ed.) *Wingspan: Inside the Men's Movement*, New York, St Martin's Press

Florio, J. (1968) *Queen Anna's New World of Words*, Menston, Scolar Press

Foster, H. (1985) *Recodings: Art, Spectacle, Cultural Politics*, Seattle, Bay Press

Foster, H. (1993) '"Primitive" scenes', *Critical Inquiry*, **20**, pp.69–102

Foster, R. (1993) *Paddy and Mr Punch*, Harmondsworth, Penguin Books

Francis, D. (1992) *The Imaginary Indian: The Image of the Indian in Canadian Culture*, Vancouver, Arsenal Pulp Press

Frankenberg, R. (1993) *White Women, Race Matters: The Social Construction of Whiteness*, Minneapolis Press, University of Minnesota

Frankenberg, R. (ed.) (1997) *Displacing Whiteness*, Durham, NC, Duke University Press

Freedland, J. (1997) 'Master race of the left', *The Guardian*, 30 August

Fryer, P. (1984) *Staying Power: The History of Black People in Britain*, London, Pluto Press

Fung, R. (1993) 'Working through cultural appropriation', *Fuse Magazine*, **16**, 5/6, pp.16–24

Füredi, F. (1998) *The Silent War: Imperialism and the Changing Perception of Race*, London, Pluto Press

Fuss, D. (1989) *Essentially Speaking: Feminism, Nature and Difference*, New York, Routledge

Gaine, C. (1987) *No Problem Here: A Practical Approach To Education and 'Race' in White Schools*, London, Hutchinson

Gallagher, C. (1995) 'White reconstruction in the university', *Socialist Review*, **24**, 1/2, pp.165–88

Galton, F. (1979) 'Hereditary talent and character', in M. Biddiss (ed.) *Images of Race*, Leicester, Leicester University Press

Gates, H., Jr (ed.) (1986) *'Race', Writing and Difference*, Chicago, University of Chicago Press

Gates, H., Jr (1988) *The Signifying Monkey*, Oxford, Oxford University Press

George, N. (1998) 'On white negroes', in D. Roediger (ed.) *Black on White: Black Writers on What it Means to be White*, Schocken Books, New York

Gidley, M. (ed.) (1992) *Representing Others: White Views of Indigenous Peoples*, Exeter, University of Exeter Press

Gilman, S. (1991) *The Jew's Body*, London, Routledge

Gilroy, B. (1987) *'There Ain't No Black in the Union Jack: The Cultural Politics of Race and Nation*, London, Macmillan

Gilroy, P. (1987) *Problems in Anti-racist Strategy*, London, Runnymede Trust

Glass, R. (1961) '"Teddy Boys" at Shepherd's Bush: an interview by Barry Carman, a B.B.C. reporter', in R. Glass *London's Newcomers: The West Indian Migrants*, Cambridge, Mass., Harvard University Press

Gobineau, A. (1915) *The Inequality of Human Races*, Volume 1, London, Heinemann

Gobineau, A. (1966) *The Inequality of Races*, Los Angeles, Noontide Press

Goldberg, H. (1976) *The Hazards of Being Male: Surviving the Myth of Masculine Privilege*, New York, Nash

Gray, R. (1981) *The Aristocracy of Labour in Nineteenth-century Britain c. 1850–1914*, London, Macmillan

Grey Owl (1931) *The Men of the Last Frontier*, London, Country Life

Grey Owl (1934) *Pilgrims of the Wild*, Toronto, Macmillan

Grey Owl (1936) *Tales of an Empty Cabin*, London, Lovat Dickson

Guillaumin, C. (1995) *Racism, Sexism, Power and Ideology*, London, Routledge

Gurnah, A. (1984) 'The politics of racism awareness training', *Critical Social Policy*, **11**, pp.6–20

Hacker, A. (1992) *Two Nations: Black and White, Separate, Hostile, Unequal*, New York, Charles Scribner's Sons

Hagan, L. (ed.) (1992) *Women Respond to the Men's Movement*, San Francisco, Pandora

Haggard, R. (1905) *The Poor and the Land*, London, Longman, Green and Co.

Haggard, R. (1976) 'Town versus country', in P. Keating (ed.), *Into Unknown England, 1866–1913: Selections from the Social Explorers*, Manchester, Manchester University Press

Haim, S. (1976) (ed.) *Arab Nationalism: An Anthology*, Berkeley, University of California Press

Hannaford, I. (1996) *Race: The History of an Idea in the West*, Washington, The Woodrow Wilson Center Press

Harding, C. (ed.) (1992) *Wingspan: Inside the Men's Movement*, New York, St Martin's Press

Harms, R. (1981) *River of Wealth, River of Sorrows: The Central Zaire Basin in the Era of the Slave and Ivory Trade, 1500–1891*, New Haven, Conn.,Yale University Press

Hay, D. (1957) *Europe: The Emergence of an Idea*, Edinburgh, Edinburgh University Press

Hayao Kawai (1998) 'Japan's self-image: what distinguishes the Japanese?' *Japan Review of International Affairs*, **12**, 2, pp.143–56

Hearn, J. (1993) 'The politics of essentialism and the analysis of the "men's movement(s)"', *Feminism and Psychology*, **3**, 3, pp.405–9

Hebdige, D. (1979) *Subculture: The Meaning of Style*, London, Methuen

Henwood, D. (1994) *The State of the U.S.A. Atlas: The Changing Face of American Life in Maps and Graphics*, London, Penguin

Hewitt, R. (1996) *Routes of Racism: The Social Basis of Racist Action*, Stoke-on-Trent, Trentham Books

Hill, M. (ed.) (1997) *Whiteness: A Critical Reader*, New York, New York University Press

Hiller, S. (ed.) (1991) *The Myth of Primitivism: Perspectives on Art*, London, Routledge

Himmelfard, G. (1971) 'Mayhew's poor: a problem of identity', *Victorian Studies*, **14**, pp.307–20

Hobsbawm, E. (1969) *Industry and Empire: From 1750 to the Present Day*, London, Pelican

Hobson, J. (1901) *The Psychology of Jingoism*, London, Grant Richards

Holinshed, R. (1965) *Holinshed's Chronicles, England, Scotland, and Ireland*, New York, AMS Press

hooks, b. (1998) 'Representations of whiteness in the black imagination', in D. Roediger (ed.) *Black on White: Black Writers on What it Means to be White*, Schocken Books, New York

Hourani, A. (1970) *Arabic Thought in the Liberal Age: 1798–1939*, Oxford, Oxford University Press

Howkins, A. (1986) 'The discovery of rural England', in R. Colls and P. Dodd (eds) *Englishness: Politics and Culture 1880–1920*, London, Croom Helm

Humphreys, A. (1977) *Travels into the Poor Man's Country: The Work of Henry Mayhew*, Firle, Caliban Books

Huxley, T. (1894) *Man's Place in Nature and Other Anthropological Essays*, London, Macmillan

Huxley, T. (1979) 'The forefathers and forerunners of the English people', in M. Biddiss (ed.) *Images of Race*, Leicester, Leicester University Press

Ignatiev, N. (1995) *How the Irish Became White*, New York, Routledge

Ignatiev, N. (1997) *The Colin Bell Show*, BBC Radio Scotland, 10 December

Inge, W. (1919) *Outspoken Essays*, London, Longman, Green and Co.

Inglehart, R. (1977) *The Silent Revolution: Changing Values and Political Styles Among Western Publics*, Princeton, NJ, Princeton University Press

Inglehart, R. (1981) 'Post-materialism in an environment of security', *The American Political Science Review*, **75**, 4, pp.880–900

ILEA (Inner London Education Authority) (1983) *Race, Sex and Class: 3, A Policy for Equality: Race*, London, Inner London Education Authority

Innes, C. (1994) 'Virgin territories and motherlands: colonial representations of Africa and Ireland', *Feminist Review*, **47**, pp.3–14

Isaacs, H. (1968) 'Group identity and political change: the role of color and physical characteristics', in J. Franklin (ed.) *Color and Race*, Boston, Houghton Mifflin

Jacobs, S. (1985) 'Race, empire and the welfare state: council housing and racism', *Critical Social Policy*, **5**, 1, pp.6–28

Jacobson, M. (1998) *Whiteness of a Different Color: European Immigrants and the Alchemy of Race*, Cambridge, Mass., Harvard University Press

Jay, E. (1992) *'Keep Them in Birmingham': Challenging Racism in South-West England*, London, CRE

Jeater, D. (1992) 'Roast beef and reggae music: the passing of whiteness', *New Formations*, **18**, pp.107–21

John, G. (1991) '"Taking sides": objectives and strategies in the development of anti-racist work in Britain', in C.D. Sterring Group (ed.) *Setting the Context for Change*, London, Central Council for Education and Training in Social Work

Jones, S. (1989) *Black Culture, White Youth: The Reggae Tradition from JA to UK*, London, Macmillan

Jordan, G. and **Weedon, C.** (1995) *Cultural Politics: Class, Gender, Race and the Postmodern World*, Oxford, Blackwell

Jordan, W. (1968) *White Over Black: American Attitudes Toward the Negro, 1550–1812*, Williamsburg, University of North Carolina Press

Joyce, J. (1987a) 'The black canon: reconstructing black American literary criticism', *New Literary History*, **18**, 2, pp.335–44

Joyce, J. (1987b) '"Who the cap fit": unconsciousness and unconscionableness in the criticism of Houston A. Baker, Jr. and Henry Louis Gates, Jr.', *New Literary History*, **18**, 2, pp.371–84

Kammer, J. (1992) '"Male" is not a four-letter word' in C. Harding (ed.) *Wingspan: Inside the Men's Movement*, New York, St Martin's Press

Kaufman, M. (1994) 'Men, feminism, and men's contradictory experiences of power', in H. Brod and M. Kaufman (eds) *Theorizing Masculinities*, Thousand Oaks, Calif., Sage

Katz, J. (1978) *White Awareness: Handbook for Anti-racism Training*, Norman, University of Oklahoma Press

Keating, P. (ed.) (1976) 'Introduction', *Into Unknown England, 1866–1913: Selections from the Social Explorers*, Manchester, Manchester University Press

Keith, A. (1922) 'The dawn of national life: an outline of racial origins: how man emerged from the horde at the call of the tribal spirit', in J. Hammeron (ed.) *Peoples of all Nations*, Volume One, London, The Amalgamated Press

Kemf, E. (ed.) (1993) *Indigenous Peoples and Protected Areas: The Law of Mother Earth*, London, Earthscan

Kimmel, M. (1995) '"Born to run": nineteenth century fantasies of masculine retreat and re-creation (or the historical rust on Iron John)', in M. Kimmel (ed.) *The Politics of Manhood: Profeminist Men Respond to the Mythopoetic Men's Movement (And the Mythopoetic Leaders Answer)*, Philadelphia, Temple University Press

Kimmel, M. and **Kaufman, M.** (1994) 'Weekend warriors: the new men's movement', in H. Brod and M. Kaufman (eds) *Theorizing Masculinities*, Thousand Oaks, Calif., Sage

Kipling, R. (1912) *Kim*, London, Macmillan

Kipnis, A. (1992) 'In quest of archetypal masculinity', in C. Harding (ed.) *Wingspan: Inside the Men's Movement*, New York, St Martin's Press

Kitahara, M. (1987) 'The Western impact on Japanese racial self-image', *Journal of Developing Societies*, **3**, pp.184–9

Knight, A. (1990) 'Racism, revolution, and indigenismo: Mexico, 1910–1940', in R. Graham (ed.) *The Idea of Race in Latin America, 1870–1940*, Austin, University of Texas Press

Kong, L. (1995) 'Folktales and reality: the social construction of race in Chinese tales', *Area*, **27**, 3, pp.261–7

Kottack, C. (1990) *Prime-Time Society: An Anthropological Analysis of Television and Culture*, Belmont, Wadsworth

Lancaster, R. (1991) 'Skin color, race, and racism in Nicaragua', *Ethnology*, **30**, 4, pp.339–53

Laurie, N. and **Bonnett, A.** (forthcoming) 'Adjusting to equity: anti-racism and the contradictions of neo-liberalism in Peru', *Antipode*

LeBon, G. (1912) *The Psychology of Peoples*, New York, G.E. Stechert

Lemann, N. (1987) 'Fake masks', *The Atlantic*, November

Lertzman, D. (1994) 'Of men and goddesses: ancient images of man and nature', *RavenFlight: A Journal of Men's Arts and Mysteries*, 1, pp.4–14

Leupp, G. (1995) 'Images of black people in late mediaeval and early modern Japan, 1543–1900', *Japan Forum*, 7, 1, pp.1–13

Levine, J. (1994) 'The heart of whiteness: dismantling the master's house', *Voice Literary Supplement*, **128**, pp.11–16

Lewis, B. (1971) *Race and Color in Islam*, New York, Harper & Row

Lewis, B. (1982) *The Muslim Discovery of Europe*, London, Weidenfeld & Nicolson

Lewis, B. (1990) *Race and Slavery in the Middle East: An Historical Inquiry*, Oxford, Oxford University Press

Lewis, L. (1995) 'Spanish ideology and the practice of inequality in the New World', in B. Bowser (ed.) *Racism and Anti-racism in World Perspective*, Thousand Oaks, Calif., Sage

Lindholm, C. (1996) *The Islamic Middle East: An Historical Anthropology*, Oxford, Blackwell

Linnaeus, C. (1956) *Caroli Linnaei Systema Naturae*, London, British Museum [facsimile of 1758 edition]

Lipsitz, G. (1998) *The Possessive Investment in Whiteness: How White People Profit from Identity Politics*, Philadelphia, Temple University Press

London Edinburgh Weekend Return Group (1979) *In and Against the State*, London, Pluto Press

Lopez, I. (1996) *White by Law: The Legal Construction of Race*, New York, New York University Press

Lorimer, D. (1978) *Colour, Class and the Victorian: English Attitudes to the Negro in the Mid-nineteenth Century*, Leicester, Leicester University Press

Low, G. (1996) *White Skins, Blacks Masks: Representation and Colonialism*, London, Routledge

Lun, S. (1975) 'On the "black-headed people"', in Yu-ning, L. (ed.) *The First Emperor of China*, White Plains, International Arts and Sciences Press

Lyde, L. (1914) *A Geography of Africa: Fifth Edition Containing Problems and Exercises*, London, Adam and Charles Black

Lyon, H. (1977) *Tenderness is Strength*, New York, Harper & Row

Macdonald, I., Bhavnani, R., Khan L. and **John G.** (1989) *Murder in the Playground: The Report of the Macdonald Inquiry into Racism and Racial Violence in Manchester Schools*, London, Longsight Press

MacGaffey, W. (1994) 'Dialogues of the deaf: Europeans on the Atlantic coast of Africa', in S. Schwartz (ed.) *Implicit Understandings: Observing, Reporting, and Reflecting on the Encounters Between Europeans and Other Peoples in the Early Modern Era*, Cambridge, Cambridge University Press

MacKenzie, J. (ed.) (1985) *Imperialism and Popular Culture*, Manchester, Manchester University Press

Madison, G. (1916) *The Passing of the Great Race; or, The Racial Basis of European History*, New York, Charles Scribner's Sons

Mailer, N. (1961) *Advertisements for Myself*, London, André Deutsch

Malcomson, S. (1991) 'Heart of whiteness: Europe goes for the globe', *Voice Literary Supplement*, March, pp.10–14

Malik, K. (1996) *The Meaning of Race*, Basingstoke, Macmillan

Marc, F. (1992) 'The "savages" of Germany', in C. Harrison and P. Wood (eds) *Art in Theory 1990–1990: An Anthology of Changing Ideas*, Oxford, Blackwell

Marsh, R. (1934) *White Indians of Darian*, New York, Putnam's

Mason, P. (1990) *Deconstructing America: Representations of the Other*, London, Routledge

Maspero, H. (1978) *China in Antiquity*, Amherst, University of Massachusetts Press

Masterman, C. (1976) 'A weird and uncanny people', in P. Keating (ed.), *Into Unknown England, 1866–1913: Selections from the Social Explorers*, Manchester, Manchester University Press

Masterman, G. (ed.) (1901) *The Heart of the Empire*, London, Fisher Urwin

Matthews, B. (1924) *The Clash of Colour*, New York, Doran

Matthews, J. (1994) 'The song of the green', *RavenFlight: A Journal of Men's Arts and Mysteries* **1**: inside back cover

Maurer, E. (1984) 'Dada and surrealism', in W. Rubin (ed.) *'Primitivism' in 20th-Century Art: Affinity of the Tribal and Modern*, New York, Museum of Modern Art

Mayhew, H. (1950) *London's Underworld: Being Selections From 'Those that Will Not Work', The Fourth Volume of 'London Labour and the London Poor'*, London, Spring Books

Mayhew, H. (1967) *London Labour and London Poor*, 4 volumes, London, Frank Cass

Mayhew, H. (1968) *London Labour and London Poor*, Volume 1, New York, Dover

Meek, J. (1994) 'Where Caucasian means black', *The Guardian*, 3 December

Miles, R. (1982) *Racism and Migrant Labour*, London, Routledge & Kegan Paul

Miles, R. (1989) *Racism*, London, Routledge

Modood, T. (1988) '"Black", Racial Equality and Asian Identity', *New Community*, **14**, 3

Modood, T. (1990a) 'British Asian Muslims and the Rushdie Affair', *Political Quarterly*, **61**, 2, pp.143–60

Modood, T. (1990b) 'Catching up with Jesse Jackson: being oppressed and being somebody', *New Community*, **17**, 1, pp.397–404

Modood, T. (1992) *Not Easy Being British: Colour, Culture and Citizenship*, Stoke-on-Trent, Trentham Books/Runnymede Trust

Moeran, B. (1996) 'The Orient strikes back: advertising and imagining Japan', *Theory, Culture and Society*, **13**, 3, pp.77–112

Money, L. (1925) *The Peril of the White*, London, W. Collins & Sons

Montaigne, M. (1993) *The Complete Essays*, Harmondsworth, Penguin

Morabia, A. (1985) 'Bernard Lewis, *Race et couleur en pays d'Islam*', *Revue de l'histoire des religions*, **202**, 3, pp.326–8

Mörner, M. (1967) *Race Mixture in the History of Latin America*, Boston, Little, Brown & Co.

Moore, R. (1992) Interview with Robert Moore, *Thunder Stick: The Journal of Vancouver M.E.N.*, **2**, 4, pp.8–11, 27

Moore, Z. (1988) 'Reflections on Blacks in contemporary Brazilian popular culture in the 1980s', *Studies in Latin American Popular Culture*, **7**, pp.213–26

Morrison, T. (1992) *Playing in the Dark: Whiteness and the Literary Imagination*, Cambridge, Mass. Harvard University Press

Morsy, S. (1994) 'Beyond the honorary "white" classification of Egyptians: societal identity in historical context', in S. Gregory and R. Sanjek (eds) *Race*, New Brunswick, Rutgers University Press

Murphy, A. (1997) *Peru Handbook*, Bath, Footprint Handbooks

Naoko, S. (1989) 'The Japanese attempt to secure racial equality in 1919', *Japan Forum*, April, pp.94–5

Nayak, A. (1997) 'Tales from the darkside: negotiating whiteness in school arenas', *International Studies in Sociology of Education*, **7**, 1, pp.57–80

Nayak, A. (1999) *Enacting Whiteness: White Youth Identities in Transition*, unpublished PhD dissertation, University of Newcastle

Nelson, A. (1990) 'Equal opportunities: dilemmas, contradictions, white men and class', *Critical Social Policy*, **10**, 1, pp.25–42

Nolde, E. (1992) 'On primitive art', in C. Harrison and P. Wood (eds) *Art in Theory 1990–1990: An Anthology of Changing Ideas*, Oxford, Blackwell

Norcross, P. (1990) 'Racial stereotyping in the all-white primary school', *Cambridge Journal of Education*, **20**, 1, pp.29–35

Nott, J. (1844) *Two Lectures on the Natural History of the Caucasian and Negro Races*, Mobile, Ala. (unpublished)

Nutini, H. (1997) 'Class and ethnicity in Mexico: somatic and racial considerations', *Ethnology*, **36**, 3, pp.227–38

Nyang, S. and **Abed-Rabbo, S.** (1984) 'Bernard Lewis and Islamic studies: an assessment', in A. Hussain, R. Olson and J. Qureshi (eds) *Orientalism, Islam and Islamists*, Brattleboro, Vt., Amana Books

Oelschlaeger, M. (1991) *The Idea of Wilderness: From Prehistory to the Age of Ecology*, New Haven, Conn., Yale University Press

Offe, C. (1984) *Contradictions of the Welfare State*, London, Hutchinson

Offe, C. (1985) *Disorganized Capitalism*, Cambridge, Polity Press

Oliart, P. (1997) Interview with Alastair Bonnett, 3 April, Lima, Catholic University of Peru, (unpublished)

Osmond, J. (1995) 'Geordie dream', *New Statesman and Society*, 24 February

Peake, M. (1985) *Gormenghast*, London, Methuen

Pelling, H. (1968) *Popular Politics and Society in Late Victorian Britain*, London, Macmillan

Phizacklea, A. and **Miles, R.** (1979) 'Working-class racist beliefs in the inner city', in R. Miles and A. Phizacklea (eds) *Racism and Political Action in Britain*, London, Routledge

Piachaud, D. (1997) 'Down but not out', *Times Literary Supplement*, 24 January, pp.3–4

Pick, D. (1988) *Faces of Degeneration: A European Disorder, c.1848–1918*, Cambridge, Cambridge University Press

Pickering, M. (1986) 'White skin black masks: 'nigger' minstrelsy in Victorian Britain', in J. Bratton (ed.) *Music Hall: Performance and Style*, Milton Keynes, Open University Press

Pierson, E. (1992) 'Creating our own event', in C. Harding (ed.) *Wingspan: Inside the Men's Movement*, New York, St Martin's Press

Pieterse, J. (1990) *Empire and Emancipation: Power and Liberation on a World Scale*, London, Pluto Press

Poliakov, L. (1974) *The Aryan Myth: A History of Racist and Nationalist Ideas in Europe*, London, Sussex University Press and Heinemann Educational Books

Powell, E. (1968) 'Speech by the Rt Hon. J. Enoch Powell M.P. to the Annual General Meeting of the West Midlands Area Conservative Political Centre', in T. Utley, *Enoch Powell: The Man and his Thinking*, London, William Kimber

Pratt, M. (1992) *Imperial Eyes: Travel Writing and Transculturation*, London, Routledge

Putnam Weale, B. (1910) *The Conflict of Colour*, London, Times

Rabbath, E. (1976) 'The common origin of the Arabs', in S. Haim (ed.) *Arab Nationalism: An Anthology*, Berkeley, University of California Press

Race Traitor: Treason to Whiteness is Loyalty to Humanity (1994) 3 (published at P.O. Box 603, Cambridge, MA 02140, U.S.A.)

Rattansi, A. (1992) 'Changing the subject? racism, culture and education', in A. Rattansi and D. Reeder (eds) *Rethinking Radical Education: Essays in Honour of Brian Simon*, London, Lawrence & Wishart

Razzell, P. (1973) *The Victorian Working Class: Selections from Letters to the Morning Chronicle*, London, Frank Cass

Reid, A. (1994) 'Early Southeast Asian categorizations of Europeans', in S. Schwartz (ed.) *Implicit Understandings: Observing, Reporting, and Reflecting on the Encounters Between Europeans and Other Peoples in the Early Modern Era*, Cambridge, Cambridge University Press

Rhodes, C. (1994) *Primitivism and Modern Art*, London, Thames & Hudson

Rich, P. (1994) *Prospero's Return? Historical Essays on Race, Culture and British Society*, London, Hansib

Roebruck, J. (1992) 'The contradictory entity of social solidarity and welfare state evolution', *Acta Sociologica*, **35**, pp.63–8

Roediger, D. (1992) *The Wages of Whiteness: Race and the Making of the American Working Class*, London, Verso

Roediger, D. (1994) *Towards the Abolition of Whiteness: Essays on Race, Politics, and Working Class History*, London, Verso

Roediger, D. (ed.) (1998a) *Black on White: Black Writers on What it Means to be White*, New York, Schocken Books

Roediger, D. (1998b) 'Plotting against Eurocentrism: the 1929 surrealist map of the world', *Race Traitor*, **9**, pp.32–9

Roediger, D. (forthcoming) '"Guineas", "wiggers" and the dramas of racialized culture', *American Literary History*

Rogin, M. (1996) *Blackface, White Noise: Jewish Immigrants in the Hollywood Melting Pot*, Berkeley, University of California Press

Rose, E., Deakin, N., Abrams, M., Jackson, V., Peston, M., Vanags, A., Cohen, B., Gaitskell, J. and **Ward, P.** (1969) *Colour and Citizenship: A Report on British Race Relations*, Oxford, Oxford University Press

Rosemont, F. (1998a) 'Notes on surrealism as a revolution against whiteness', *Race Traitor*, **9**, pp.19–31

Rosemont, F. (1998b) 'Surrealists on whiteness from 1925 to the present', *Race Traitor*, **9**, pp.5–18

Ross, A. and **Rose, T.** (eds) (1994) *Microphone Fiends: Youth Music and Youth Culture*, London, Routledge

Rotundo, E. (1993) *American Manhood: Transformations in Masculinity from the Revolution to the Modern Era*, New York, Basic Books

Rubin, W. (ed.) (1984) *'Primitivism' in 20th-Century Art: Affinity of the Tribal and Modern*, New York, Museum of Modern Art

Rubio, P. (1993) 'The "exceptional white" in popular culture', *Race Traitor*, **2**, pp.68–80

Rubio, P. (1994) 'Phil Rubio replies', *Race Traitor*, **3**, pp.124–5

Ruggles, P. and **O'Higgens, M.** (1987) 'Retrenchment and the New Right', in M. Rein, G. Esping-Anderson and L. Rainwater (eds) *Stagnation and Renewal in Social Policy*, New York, M.E. Sharpe

Runnymede Trust (1991) *Race Issues Opinion Survey*, London, Runnymede Trust

Sacks, K. (1994) 'How did Jews become white folks?', in S. Gregory and R. Sanjek (eds) *Race*, New Brunswick, Rutgers University Press

Sakamoto, R. (1996) 'Japan, hybridity and the creation of colonialist discourse', *Theory, Culture and Society*, **13**, 3, pp.113–28

Sartwell, C. (1998) *Act Like You Know: African-American Autobiography and White Identity*, New York, Routledge

Sato, K. (1997) '"Same language, same race": The dilemma of Kabun in modern Japan', in F. Dikötter (ed.) *The Construction of Racial Identities in China and Japan*, London, Hurst

Sautman, B. (1994) 'Anti-black racism in post-Mao China', *The China Quarterly*, **138**, pp.411–37

Saynor, J. (1995) 'Living in precarious times', *The Observer*, 27 August

Schafer, E. (1963) *The Golden Peaches of Samarkand: A Study of T'ang Exotics*, Berkeley, University of California Press

Schwalbe, M. (1995) *Unlocking the Iron Cage: Understanding the Mythopoetic Men's Movement*, Oxford, Oxford University Press

Segal, D. (1991) '"The European": allegories of racial purity', *Anthropology Today*, **7**, 1, pp.7–9

Semmell, B. (1960) *Imperialism and Social Reform*, London, Allen & Unwin

Senghor, L. (1997) 'Negritude: a humanism of the twentieth century', in R. Grinker and C. Steiner (eds) *Perspectives on Africa: A Reader in Culture, History, and Representation*, Oxford, Blackwell

Seton, E. (1903) *Two Little Savages*, New York, Doubleday, Page & Co.

Sharma, S. (1987) '"Education for All" on Wheels: the Intercultural Mobile Unit', *Multicultural Teaching*, **5**, 2, pp.13–16

Sherard, R. (1897) *The White Slaves of England: Being True Pictures of Certain Social Conditions in the Kingdom of England in the Year 1897*, London, J. Bowden

Sherman, A. (1965) 'Deptford', in N. Deakin (ed.) *Colour and the British Electorate, 1964*, London, Pall Mall Press

Short, J. (1991) *Imagined Country: Society, Culture and Environment*, London, Routledge

Showalter, E. (1992) *Sexual Anarchy*, London, Virago

Siddle, R. (1997) 'Ainu: Japan's indigenous people', in M. Weiner (ed.) *Japan's Minorities: The Illusion of Homogeneity*, London, Routledge

Simpson, A. (1993) *Xuxa: The Mega-Marketing of Gender, Race, and Modernity*, Philadelphia, Temple University Press

Sims, G. (1976) 'The dark side of life', in P. Keating (ed.) *Into Unknown England, 1866–1913: Selections from the Social Explorers*, Manchester, Manchester University Press

Sivanadan, A. (1985) 'RAT and the degradation of Black struggle', *Race and Class*, **26**, 4, pp.1–33

Skidmore, T. (1974) *Black into White: Race and Nationality in Brazilian Thought*, New York, Oxford University Press

Skidmore, T. (1985) 'Race and class in Brazil: historical perspectives', in P. Fontaine (ed.) *Race, Class and Power in Brazil*, Los Angeles, Center for Afro-American Studies, University of California

Skidmore, T. (1990) 'Racial ideas and social policy in Brazil, 1870–1940', in R. Graham (ed.) *The Idea of Race in Latin America, 1870–1940*, Austin, University of Texas Press

Smith A. (1910) *The Wealth of Nations*, London, J.M. Dent

Smith, C. (1848) *The Natural History of the Human Species, its Typical Forms, Primaeval Distribution, Filiations, and Migrations*, Edinburgh, Lizars

Smith, D. (1990) *From the Land of Shadows: The Making of Grey Owl*, Saskatoon, Western Producer Prairie Books

Snowden, F. (1983) *Before Color Prejudice: the Ancient View of Blacks*, Cambridge, Mass., Harvard University Press

Spencer, J. (1995) 'Occidentalism in the East: the uses of the West in the politics and anthropology of South Asia', in J. Carrier (ed.) *Occidentalism: Images of the West*, Oxford, Oxford University Press

Spivak, G. (1990) 'Gayatri Spivak on the politics of the subaltern', *Socialist Review*, **20**, 3, pp.81–97

Spurr, D. (1993) *The Rhetoric of Empire: Colonial Discourse in Journalism, Travel Writing, and Imperial Administration*, Durham, NC, Duke University Press

Stafford, C. (1993) *'The Discourse of Race in Modern China:* book review', *Man: The Journal of the Royal Anthropological Institute*, **28**, 3, p.609

Starhawk (1982) *Dreaming the Dark: Magic, Sex and Politics*, Boston, Beacon Press

Stedman Jones, C. (1971) *Outcast London: A Study in the Relationship Between Classes in Victorian Society*, Harmondsworth, Penguin

Stelarc and **Linz, R.** (1996) 'Voltage in/voltage out', performance at Zone Gallery, Newcastle

Steiner, C. (1997) 'The invisible face: masks, ethnicity, and the state in Côte d'Ivoire', in R. Grinker and C. Steiner (eds) *Perspectives on Africa: A Reader in Culture, History and Representation*, Oxford, Blackwell

Stoddard, L. (1920) *The Rising Tide of Color against White World Supremacy*, New York, Charles Scribner's Sons

Stoker, B. (1897) *Dracula*, London, A. Constable & Co.

Stothers, R. (1992) 'Questioning the movement that gave me my life back', *Thunder Stick: The Journal of Vancouver M.E.N.*, **2**, 4, pp.5, 20–21

Stowe, H.B. (1852) *Uncle Tom's Cabin; or, Life Among the Lowly*, Boston, J.P. Jewett & Co.

Street, B. (1975) *The Savage in Literature*, London, Routledge & Kegan Paul

Sullivan, M. (1994) 'The 1988–1989 Nanjing anti-African protests: racial nationalism or national racism?' *The China Quarterly*, **138**, pp.438–57

Taussig, M. (1993) *Mimesis and Alterity: A Particular History of the Senses*, New York, Routledge

Taylor, B. (1984) 'Multicultural education in a "monocultural region', *New Community*, **12**, 1, pp.1–8

Taylor, B. (1986) 'Anti-racist education in non-contact areas: the need for a gentle approach', *New Community*, **13**, 2, pp.177–84

Taylor, B. (1987) 'Anti-racist education in predominantly white areas', *Journal of Further and Higher Education*, **11**, 3, pp.45–61

Taylor, B. (1988) 'Anti-racist education: theory and practice (a review article)', *New Community*, **14**, 3, pp.481–5

Taylor, B. (1990) 'Multi-cultural education in the "white highlands" after the 1988 Education Reform Act', *New Community*, **16**, 3, pp.369–78

Taylor, Q. (1990) 'Blacks and Asians in a white city: Japanese Americans and African Americans in Seattle, 1890–1940', *Western Historical Quarterly*, **23**, 4, pp.545–90

Taylor-Gooby, P. (1989) 'Current developments in the sociology of welfare', *The British Journal of Sociology*, **40**, 4, pp.637–56

Thompson, B. and **Tyagi, S.** (eds) (1996) *Names We Call Home: Autobiography on Racial Identity*, Routledge, New York

Thompson, K. (1991) 'A man needs a lodge', in K. Thompson (ed.) *To Be A Man: In Search of the Deep Masculine*, New York, Jeremy P. Tarcher/ Perigee Books

Thoreau, H. (1906) *The Writings of Henry David Thoreau*, Volume Five: *Excursions and Poems*, Boston, Houghton Mifflin

Thoreau, H. (1962) *Walden and Other Writings by Henry David Thoreau*, New York, Bantam

Thoreau, H. (1987) *Maine Woods*, New York, Harper & Row

Todorov, T. (1993) *On Human Diversity: Nationalism, Racism, and Exoticism in French Thought*, Cambridge, Mass., Harvard University Press

Torgovnick, M. (1990) *Gone Primitive: Savage Intellects, Modern Lives*, Chicago, University of Chicago Press

Troyna, B. and **Hatcher, R.** (1992) *Racism in Children's Lives: A Study of Mainly White Primary Schools*, London, Routledge

Tsurumi, S. (1987) *A Cultural History of Postwar Japan 1945–1980*, London, KPI

Twine, F. (1998) *Racism in a Racial Democracy: The Maintenance of White Supremacy in Brazil*, New Brunswick, Rutgers University Press

Tzara, T. (1992) *Seven Dada Manifestos and Lampisteries*, Calder Press, London

Vasconcelos, J. (1997) *The Cosmic Race: A Bilingual Edition*, Baltimore, Johns Hopkins University Press

Vaughan, A. (1995) *Roots of American Racism: Essays on the Colonial Experience*, New York, Oxford University Press

Verstegan, R. (1605) *A Restitution of Decayed Intelligence in Antiquities Concerning the Most Noble and Renowned English Nation*, Antwerp, R. Brunley

Wagatsuma, H. (1968) 'The social perceptions of skin color in Japan', in J. Franklin (ed.) *Color and Race*, Boston, Houghton Mifflin

Walker, D. (1998) 'Whites as heathens and Christians', in D. Roediger (ed.) *Black on White: Black Writers on What it Means to be White*, New York, Schocken Books

Ware, V. (1992) *Beyond the Pale: White Women, Racism and History*, London, Verso

Watts, H. (1988) 'The Dada event: from transsubstantiation to bones and barking', in S. Foster (ed.) *'Event' Arts and Art Events*, Ann Arbor, Mich, UMI Research Press

Weiner, B. (1992) *Boy Into Man: A Fathers' Guide to Initiation of Teenage Sons*, San Francisco, Transformation Press

Wellman, D. (1977) *Portraits of White Racism*, Cambridge, Cambridge University Press

Wiedemann, F. (1991) 'In search of a modern male initiation', in K. Thompson (ed.) *To Be A Man: In Search of the Deep Masculine*, New York, Jeremy P. Tarcher/Perigee Books

Williams, A. (1981) *Life in a Railway Factory*, New York, Garland Press

Willis, F. (1948) *101 Jubilee Street: A Book of London Yesterdays*, London, Phoenix House

Wilson, D. (1984a) 'A paler shade of yellow', *New Society*, 11 October, pp.50–53

Wilson, D. (1984b) 'Black and white view from the Middle Kingdom', *Far Eastern Economic Review*, 13 December, pp.52–5

Wolf-Light, P. (1993) 'Wild at heart', *Achilles Heel: The Radical Men's Magazine*, **15**, pp.22–4

Wolf-Light, P. (1994) 'The shadow of Iron John', *Achilles Heel: The Radical Men's Magazine*, **17**, pp.14–17

Woolcott, D. (1965) 'Southall', in N. Deakin (ed.) *Colour and the British Electorate*, 1964, London, Pall Mall Press

Worsley, P. (1964) *The Third World*, London, Weidenfeld & Nicolson

Worsley, P. (1984) *The Three Worlds*, London, Weidenfeld & Nicolson

Wray, M. and **Newitz, A.** (eds) (1997) *White Trash: Race and Class in America*, New York, Routledge

Wright, H. (1994) 'Like flies in buttermilk: Afrocentric students in the multi-cultured classroom', *Orbit*, **25**, 2, pp.29–37

Wright, W. (1990) *Café con leche: Race, Class and National Image in Venezuela*, Austin, University of Texas Press

X, Malcolm (1987) *Two Speeches by Malcolm X*, New York, Pathfinder

Young, M. (1988) Review of "anti racist science teaching"', *Multicultural Teaching*, **6**, 3, p.40

Young, R. (1995) *Colonial Desire: Hybridity in Theory, Culture and Race*, London, Routledge

Ziadet, A. (1986) *Western Science in the Arab World: The Impact of Darwinism, 1860–1930*, London, Macmillan

Index

Index

Verstegan, R., 17
Vink, N., 64

Wagatsuma, H., 27, 69, 74
Waitz, T., 21
Ware, V., 6, 23–24
Watts, H., 83
Weiner, B., 108–110
welfare state, 30, 39–45, 132–33
Western identity, 4, ch.3 *passim*, 78, 80, 82, 84–9, 95, 103, 110–15, 138, 142
White, C., 17–18
whitening, policy and practice, 5, 50–61, 75
'White studies', 1–3, 6, 44–5, 119–120, 124–141
'white trash', 136

'wiggers' 136
wilderness, 51, 56, 79, 80, 89–110, 116
Williams, A., 39
Willis, F., 39
Wolf-Light, P., 104–5
women, 18, 22–27, 39, 46–8, 52, 62–7, 72–5, 97, 101, 104, 107–8, 124, 135
Wright, W., 55–60, 76

Xuxa, 5, 49, 62–67

Young, R., 21, 136

Zaire, 14
Zhang X., 10–11